Free Energy Transduction in Biology

The Steady-State Kinetic and Thermodynamic Formalism

FREE ENERGY TRANSDUCTION IN BIOLOGY

The Steady-State Kinetic and Thermodynamic Formalism

Terrell L. Hill
NATIONAL INSTITUTES OF HEALTH
BETHESDA, MARYLAND

ACADEMIC PRESS New York San Francisco London 1977

A Subsidiary of Harcourt Brace Jovanovich, Publishers

ACADEMIC PRESS, INC.
111 Fifth Avenue, New York, New York 10003

United Kingdom Edition published by
ACADEMIC PRESS, INC. (LONDON) LTD.
24/28 Oval Road, London NW1

Library of Congress Cataloging in Publication Data

Hill, Terrell L
Free energy transduction in biology.

Includes bibliographical references.
1. Bioenergetics. 2. Thermodynamics.
I. Title.
QH510.H54 574.1′9121 77–1228
ISBN 0–12–348250–X

PRINTED IN THE UNITED STATES OF AMERICA

To the memory of
John G. Kirkwood,
pioneer in theoretical molecular biology

Contents

Preface

The subtitle of this book is very important. The book does not cover the entire field of free energy transduction in biology but rather only one special topic: the steady-state kinetic and thermodynamic formalism related to free energy transduction. As the word "formalism" implies, the discussion concerns general principles and methods and not details of proposed mechanisms in the various special cases. Although the main argument is put forward in terms of examples, the examples are chosen, for the most part, for pedagogical reasons rather than as serious models.

The advantage of this kind of approach is that one can attain a kind of overview of how free energy transduction is accomplished. The disadvantage is that most current research activity quite naturally lies elsewhere: in attempts to establish the molecular mechanism in particular cases. But as the latter work progresses and detailed models are suggested, the formalism presented here should prove useful in the calculation and understanding of the steady-state kinetic and thermodynamic properties implicit in such models.

Perhaps the main theme of the present book is that, with respect to general principles, free energy transduction can be quite simply understood in terms of conventional kinetics and thermodynamics—suitably related to each other. In particular, it will be argued that free energy transduction is accomplished by complete biochemical cycles and not by individual steps or transitions (as often assumed). A brief abstract of this argument is to be published in *Trends in Biochemical Sciences* (1977).

The book attempts to bring together, in a single coherent account, work published over the past ten years in various research papers. In addition, many new examples and much new material have been added.

Additional introductory comments are included in the first section of Chapter 1.

The writing of this book was greatly aided by the skill and patience of Mrs. Alma Martinson, who typed the manuscript.

I am indebted, for very valuable comments on the manuscript or its subject matter, to Drs. Britton Chance, Elliott Charney, Don DeVault, John Gergely, Joel Keizer, David Kliger, Robert Simmons, Eugene Switkes, and Peter von Hippel.

Note on Notation

The symbol A' was originally introduced for the "basic free energy" [*Progr. Biophys. Mol. Biol.* **28**, 267 (1974)] because of its relationship to the canonical partition function Q. This partition function is a practical one to use in macromolecular statistical mechanics, whereas the partition function Δ (related to the Gibbs free energy G) is not. However, for formal thermodynamic and kinetic purposes, as in this book, one might as well use G' instead of A' because G' is the exact quantity for a constant pressure system while A' is a close approximation of it. The reader should feel free, if he wishes, to make a mental substitution of G' for A' throughout the book.

Chapter 1 The Diagram Method: States

1.1 Introduction

There are many examples in biochemistry in which a macromolecule, usually an enzyme or enzyme complex, can exist in a finite number of discrete states and such that the macromolecule undergoes continuous cycling among these states at steady state. Ligands, substrates, and products (at fixed concentrations) are involved in some of the transitions between states, but in the formal kinetics the macromolecule plays the central role because it is present in every state. Although simple steady conversion of substrate into product by an enzyme is an example, the more interesting cases involve transduction of one form of free energy into another, as in various kinds of active transport, oxidative phosphorylation, phototranslocation coupling, muscle contraction, etc. Analysis of systems of this type provides a foundation for the understanding of the general principles involved in many, if not most, bioenergetic transformation problems.

Let us try to be quite clear at the outset about the level and the generality of the theory to be summarized in this book. What will be presented here is *not* a theory at the most fundamental molecular or atomic level. In fact, at the present time there is no single example of free energy transduction about which sufficiently detailed experimental information is available to allow the construction (with confidence) of a complete molecular model or theory. Such models can be expected to come along, one at a time, in the future. [Because of its relative simplicity, phototranslocation coupling (1, 2) in the purple membrane of *Halobacterium halobium* is likely to provide the first example.] In fact, we can never expect a completely *general* theory of free energy transduction at the *molecular–atomic* level because of the vast variety of detail that must be encompassed. Rather, molecular models must be approached on an ad hoc basis, though undoubtedly many of them will prove to be closely related to one another.

On the other hand, in this book, we do not go to the opposite extreme of presenting a discussion of free energy transduction in completely phenomenological terms—for example, in the language of Onsager's nonequilibrium thermodynamics suitably generalized, as would be required for most of these problems, to apply very far from equilibrium.

Instead, we follow an intermediate course. We deal with macromolecules (proteins, enzymes, complexes), their interstate transitions (including the smaller molecules with which they interact), and the rate constants governing the probabilities of these transitions. But, as suggested above, we do not attempt, in any example, to furnish an ab initio theory of the rate constants per se. We take the macromolecular states, transitions, and rate constants as *given*, and then examine the nature of free energy transduction, and a number of related topics, in these terms and at this level of detail. This approach seems to provide the clearest possible overview of the *general* theoretical principles involved in free energy transduction; it does not lose sight of the forest for the trees, as must almost inevitably be the case in a completely detailed molecular analysis of various special cases. Thus the level we adopt permits of very considerable generality and allows further details to be incorporated into particular examples, as the details become available, without disturbing the validity of the general kinetic formalism to be presented here.

Incidentally, exactly this same level of detail—neither ab initio molecular nor phenomenological—is used with great effect by Joel Keizer (3) in his recent very general treatment of dissipation and fluctuations in far-from-equilibrium thermodynamic systems.

The analysis of steady-state enzyme action and free energy transduction in terms of the rate constants operating between discrete macromolecular states is, of course, routine rather than novel in special cases (1, 4–14). But our object in the present monograph is to present a single systematic formalism that is applicable to a wide variety of such examples. We are interested here in analytical methodology rather than in particular models for particular problems. This same point of view was adopted, incidentally, in two recent papers (15, 16) on the theory of muscle contraction; but, of course, experimentally founded "particular models" are the *ultimate* goal.

The main analytical tool in the study of these discrete-state, cycling systems is a diagram method introduced by King and Altman (17) in 1956 and rediscovered and extended in several ways (cycles, cycle fluxes, coupling, reciprocal relations, membrane transport) by Hill (18, 19) in 1966. In fact, it is the extensions of the King–Altman method that will be found of most use in the present book. More recently, the different kinds of free energy levels of the macromolecular states, especially at steady state, have been discussed by Hill (15, 20), Hill and Simmons (21, 22), and Simmons and Hill (23). Also, fluctuations and noise in the steady-state probabilities of states and in cycle fluxes have been investigated by Hill (20, 24), Hill and Chen (25), Chen and Hill (26), and Chen (27–29). Finally, multienzyme complexes have received some attention (30). Thus, a substantial theoretical foundation is now available as an aid in the study and understanding of these systems.

Although the object of this book is to provide a unified account of these subjects, we shall use, for this purpose, illustrative examples rather than abstract generalities as much as possible. This should make much of the material here—especially the essentials—easily accessible to nontheoretical biochemists and biophysicists. Furthermore, most of the numerous examples will be chosen strictly for their pedagogical value rather than as models to be taken seriously. One aim is to provide sufficient and suitable examples so that the interested reader will be able to analyze his own models by these methods.

Although expressed in quite different language (18, 19), Chapter 1 has for its foundation the King–Altman (17) diagram method for the calculation of steady-state probabilities of states. Chapter 2 then introduces the essential topic of cycles, cycle fluxes, etc. Chapters 3–5 contain a discussion of the more important bioenergetic principles that emerge from the diagram approach. These are the most important chapters in the book. Chapters 6 and 7 are concerned with somewhat more specialized aspects of the subject: stochastics and fluctuations (Chapter 6); and interacting subsystems and multienzyme complexes, including oxidative phosphorylation (Chapter 7). Incidentally, Chapter 7 does not depend at all on Chapters 5 and 6.

Certain important special topics are treated briefly in the appendices: "reduction" of diagrams; membrane potential and charge carriers; and systems that make use of photon absorption. As will be explained in Appendix 5, systems that absorb (or produce) radiant energy are exceptional and do not fully fit into the formalism of this book: Chapters 1, 2, and 6 are applicable to such systems, but not Chapters 3 and 4 as they stand.

A chapter on noise theory, which would be a logical extension of Chapter 6, is omitted because it would have to be relatively sophisticated mathematically, and because a review of this topic has just been written by Chen (31).

Actually, much of the discussion in this book (all but Chapters 4 and 5) is more general than implied above. That is, some of the analysis applies to *any* first-order discrete-state kinetic system at steady state (19, 32). However, for definiteness and because the motivation here is biochemical, we shall introduce and maintain a macromolecular or enzymatic context throughout.

1.2 Diagrams for Steady-State (and Equilibrium) Systems

We consider a large number N (an ensemble) of equivalent and independent macromolecular units or systems (e.g., one unit equals one enzyme molecule or complex), either free in solution or immobilized (e.g., in a membrane or in a myofilament). Each unit may exist in any one of n discrete states, $i = 1, 2, \ldots, n$. Transitions are possible between some pairs of these states (possibly all pairs). Ligands, substrates, and products may be involved in some transitions but, if they are, their concentrations are assumed to be essentially constant over the time scale of interest here. All transitions are treated as first-order processes; the first-order rate constant for the transition $i \to j$ is denoted by α_{ij}. For example, for the binding transition E (enzyme) + S (substrate) → ES, we would use $\alpha = \alpha^* c_S$, where α is the first-order rate constant, α^* the second-order rate constant, and c_S the concentration of substrate (see Section 3.1 for further details).

The n possible states for each unit can be represented by points in a diagram, with a line between two points indicating possible inverse transitions. For example, Fig. 1.1b is the diagram representing the kinetic scheme in Fig. 1.1a (where P means product).

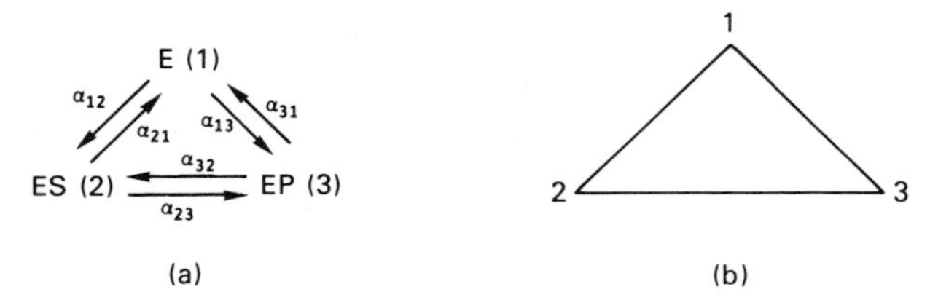

FIG. 1.1 (a) First-order rate constant notation. (b) "Diagram" corresponding to (a).

A few other examples of diagrams are shown in Fig. 1.2 (many others will be encountered later in the book). Figure 1.2a could represent a condensation of Fig. 1.1b if state 3 in Fig. 1.1b is a transient intermediate (see Appendix 1). Another possibility would be: state 1 = E in a membrane; state 2 = EL in the membrane; L is a ligand present in both baths (A and B), on either side of the membrane (usually $c_A \neq c_B$), with binding of L on E possible from either bath. The left-hand line in Fig. 1.2a would then represent, say, transitions involving adsorption–desorption from bath A while the right-hand line relates in the same way to bath B.

Figure 1.2b is the diagram of a common type of system, one with consecutive reactions. For example, the Hodgkin–Huxley potassium channel (four subunits) in the squid giant axon membrane has this kind of diagram but with five consecutive states (33).

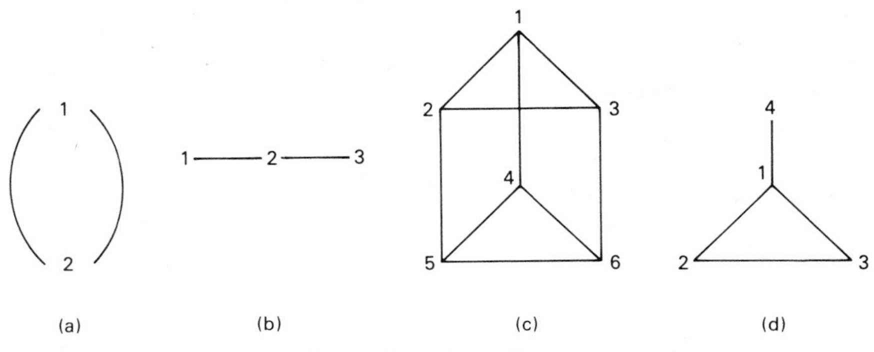

FIG. 1.2 Examples of diagrams.

Figure 1.2c might represent an expansion of Fig. 1.1b if the enzyme can exist in two conformations, E and E*, with state 4 = E*, state 5 = E*S, and state 6 = E*P. The same diagrams would obtain if E binds, on a separate site, a ligand L that may or may not modify the enzyme kinetics. In this case, 4 = LE, 5 = LES, and 6 = LEP. An example of this would be E = myosin, S = ATP, P = ADP + P_i, and L = actin. Figure 1.2d is a special case of Fig. 1.2c in which states 5 and 6 are unstable (i.e., S and P do not bind appreciably to E* or to LE in the two examples above).

If $N_i(t)$ is the number of units of the ensemble in state i at t, the kinetic equations for the diagram in Fig. 1.1b (as an example) are

$$dN_1/dt = (\alpha_{21}N_2 - \alpha_{12}N_1) + (\alpha_{31}N_3 - \alpha_{13}N_1), \tag{1.1}$$

with similar relations for dN_2/dt and dN_3/dt, together with the conservation relation

$$N_1 + N_2 + N_3 = N. \tag{1.2}$$

However, one of the three differential equations is redundant (i.e., not independent). In any example the differential equations are automatically implied by the diagram: each line in the diagram leading into state i will provide a pair of terms in the expression for dN_i/dt, as in Eq. 1.1.

The probability of state i, that is, the fraction of units in state i, is $p_i = N_i/N$. Fluctuations in the N_i, and related topics, will not be considered until Chapter 6.

At $t = \infty$, all $dN_i/dt = 0$ and the ensemble of N units will be either at equilibrium or at a nonequilibrium steady state. A steady state is possible only if the diagram contains at least one cycle (a single closed path in the diagram, not including any appendages). Thus Fig. 1.2b, with no cycle, can only lead to equilibrium at $t = \infty$. The other diagrams in Figs. 1.1 and 1.2 contain cycles (indeed, Fig. 1.2c has 14 different cycles—see Fig. 4.13). If a diagram possesses one or more cycles, the rate constants *might* have values such that the ensemble reaches equilibrium at $t = \infty$, but in general a steady state is to be expected. If the product of rate constants in a particular direction around any given cycle κ of a diagram is designated by $\Pi_{\kappa+}$, and the product in the opposite direction is designated $\Pi_{\kappa-}$, then the condition for equilibrium is $\Pi_{\kappa+} = \Pi_{\kappa-}$ for every cycle in the diagram. (Ordinarily we take + to be counterclockwise for each cycle.) For example, in Fig. 1.1a, the condition is $\alpha_{12}\alpha_{23}\alpha_{31} = \alpha_{21}\alpha_{32}\alpha_{13}$. This requirement is a straightforward consequence of the application of detailed balance at equilibrium (e.g., $\alpha_{21} N_2^e = \alpha_{12} N_1^e$ in Fig. 1.1a and Eq. 1.1) to each line in the cycle being considered.

The individual transitions in each of the N systems (units) of an ensemble are stochastic in nature. Therefore, in a collection of identically prepared ensembles, or if the same experiment is repeated over and over on a single ensemble, we would encounter fluctuations in the quantities $N_i(t)$ about mean values $\bar{N}_i(t)$. It is actually the mean values that appear in equations such as 1.1. But, for simplicity of notation, we shall omit mean value overbars on the N_i (and on the fluxes, below) until needed explicitly in Chapter 6.

1.3 Directional Diagrams and the Steady-State Populations of States

We shall introduce this subject by means of a hypothetical model for active transport of one ligand by another across a membrane. As indicated in Fig. 1.3a, a protein E has a site (·) for binding a ligand L_1 and another site (×) for binding a second ligand L_2. Both baths contain both ligands at the concentrations indicated in the figure. However, the L_2 site on E is "activated" only if L_1 is already bound. The protein, with ligand L_1 bound, can undergo a conformational change ($2 \rightleftarrows 3$ or $4 \rightleftarrows 5$) that has the effect of switching the bath to which the binding sites are accessible. The diagram is shown in Fig. 1.3b. It can be thought of, for example, as a reduced form of the diagram in Fig. 1.3c, if there is a fast equilibrium between states 1 and 1′ (i.e., a fast conformational change in E in the absence of ligands); see Appendix 1 in this connection.

As the system (i.e., $E + L_1 + L_2$) moves via transitions around the diagram (Fig. 1.3b), completing cycles of the three types possible, the net effect is to transport L_1 and L_2 from one bath to the other. At equilibrium $c_{1A} = c_{1B}$, $c_{2A} = c_{2B}$, and there is no average net transport of either ligand. But at steady state, where we have concentration inequalities rather than equalities, a sufficient concentration difference in one ligand can cause a net flux in the other ligand *against* its own concentration gradient. That is, a free

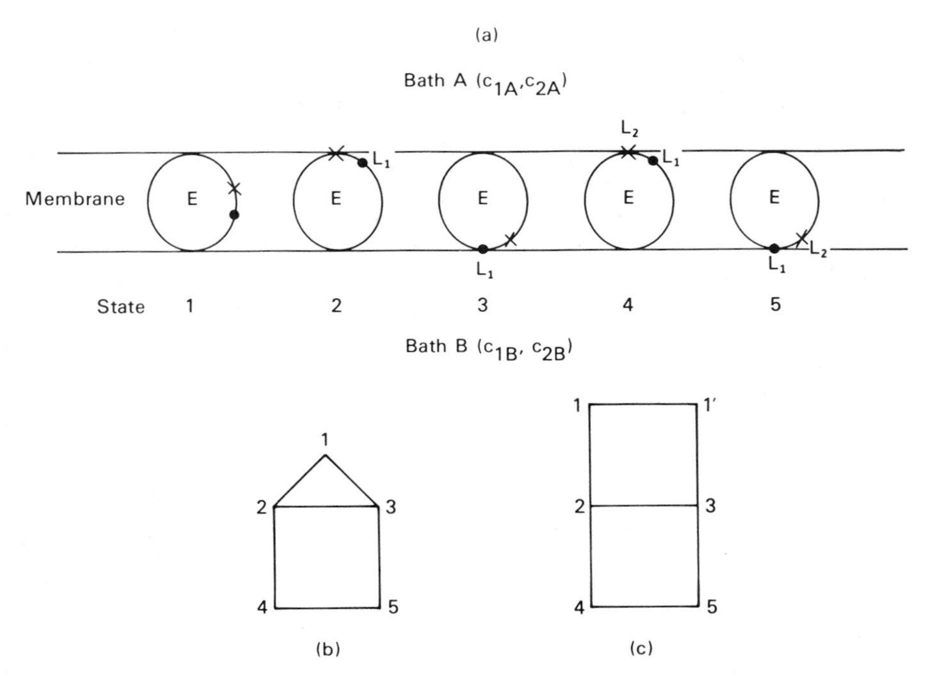

FIG. 1.3 (a) Illustrative model for membrane transport of two ligands L_1 and L_2 between two baths A and B: •, site for L_1; ×, site for L_2; E, macromolecule or enzyme. (b) Diagram corresponding to (a). (c) Expanded diagram if state 1 is expanded into two states.

energy decrease in one ligand can be partially converted into a free energy increase in the other, with a certain nonzero efficiency. This aspect of the model is pursued in Chapter 3. But we turn now to the algebraic problem.

Our main concern is with steady-state fluxes, but we are interested also in the steady-state values of the N_i, denoted by N_i^∞ (equilibrium values of the N_i will be indicated by N_i^e). Of course, the N_i^∞ may be found in a straightforward way, in any particular case, by solution of a set of linear algebraic equations (see Eq. 1.4, below). But if the model is at all complicated, this may involve a great deal of tedious labor. One of our main objects in this chapter is to show how the solution of the algebraic equations can be found, alternatively, from an enumeration of a certain class of diagrams (17–19). Furthermore, the solution in terms of diagrams has a certain intuitive appeal, and leads directly to the net flux between any transition-pair of states in the diagram.

In the above paragraph, we are referring to a solution of the linear equations for the N_i^∞ as *explicit functions* of all the rate constants of the model and N. Of course if one needs only *numerical* solutions for the N_i^∞ or $p_i^\infty = N_i^\infty/N$ in particular cases, the job is most simply done by computer without reference to diagrams.

The directional diagrams (17–19) introduced in this chapter, then, are valuable in providing explicit solutions for the p_i^∞. But the flux diagrams (18, 19) defined in Chapter 2 play a more fundamental role: they furnish the basis for a comprehension of the various *components* of flux present in a steady-state system with a multicycle diagram, and therefore of such properties as thermodynamic "coupling," free energy transfer, reciprocal relations (near equilibrium), rate of entropy production, fluctuations and noise in fluxes, etc.

The differential equation for N_1 is

$$\frac{dN_1}{dt} = (\alpha_{21} N_2 - \alpha_{12} N_1) + (\alpha_{31} N_3 - \alpha_{13} N_1), \tag{1.3}$$

with similar expressions for dN_2/dt, etc. Each pair of terms on the right corresponds to a line in the diagram, Fig. 1.3b. Thus dN_2/dt is equal to three pairs of terms, and there are three lines emanating from state 2 in the diagram; etc. At steady state, Eq. 1.3 becomes

$$(\alpha_{21} N_2^\infty - \alpha_{12} N_1^\infty) + (\alpha_{31} N_3^\infty - \alpha_{13} N_1^\infty) = 0. \tag{1.4}$$

At equilibrium, each pair of terms is *separately* equal to zero (detailed balance).

We obtain an equation like Eq. 1.4 for each state. Thus, in this example, we have a set of five linear equations in the five N_i^∞. But only four of these equations are independent. The fifth independent equation, necessary to solve for the N_i^∞, is $\sum_i N_i^\infty = N$. The solution will give each $p_i^\infty = N_i^\infty/N$ as a function of rate constants.

However, instead of solving for five unknowns in the conventional way, as an alternative we can write the solution using diagrams as follows. [The proof (18) is given in Appendix 2.] The first step is to construct the complete set of *partial diagrams*, each of which contains the maximum possible number of lines (four here) that can be included in the diagram without forming any cycle (closed path). There are eleven such partial diagrams in this case, shown in Fig. 1.4.

If one more line is introduced into any vacant position in any of these partial diagrams, a cycle is produced.

At least one line goes to each vertex (state) in a partial diagram (otherwise more lines could be introduced without forming a cycle).

The next step is to introduce arrows (i.e., a directionality for each line) into the partial diagrams of Fig. 1.4 in five different ways, one way for each state (vertex). For example, consider state 2. Figure 1.5 shows the eleven

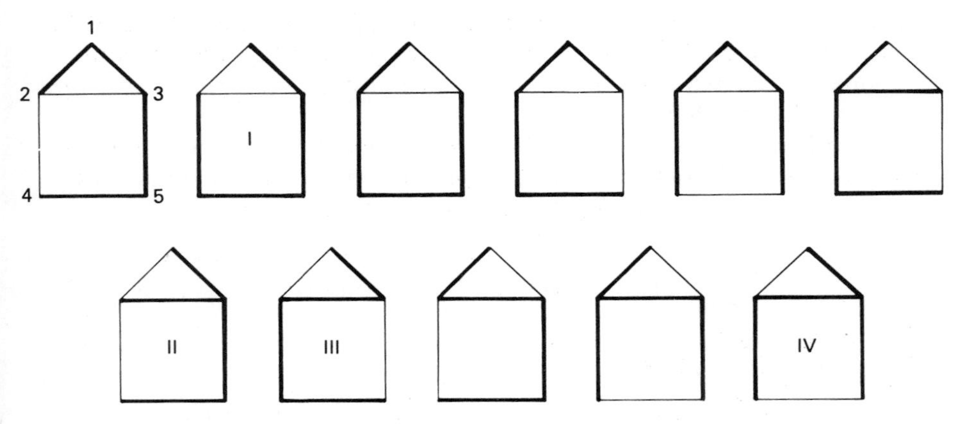

FIG. 1.4. Partial diagrams for Fig. 1.3b. I, II, III, and IV are referred to in text (Chapter 2).

directional diagrams for this state, as obtained from Fig. 1.4. The recipe for introducing arrows is simple: all connected paths in Fig. 1.5 are made to "flow" *toward* and *end at* vertex 2. It will be noted that in the flow toward the ultimate vertex (vertex 2 in Fig. 1.5), "streams" may converge but they never diverge (for this would require a cycle in the partial diagram).

There is a set of eleven directional diagrams for each of the five states. In each case, all streams flow toward—and end at—the particular state being considered.

Now each directional line or arrow in Fig. 1.5 corresponds to a rate constant; the key for assigning rate constants to directional lines is provided

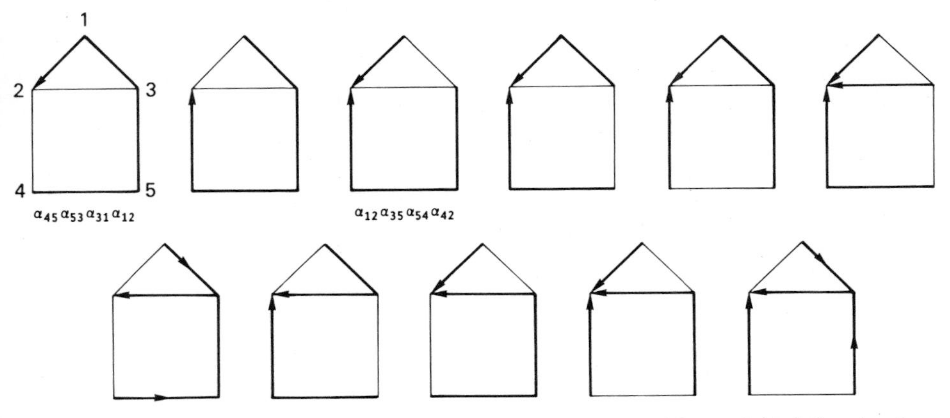

FIG. 1.5 Directional diagrams for state 2. Algebraic values of first and third directional diagrams are given.

by the kinetic scheme corresponding to the diagram (e.g., compare Figs. 1.1a and 1.1b). Using this key, each directional diagram in Fig. 1.5 represents a product of four rate constants, one for each arrow, as indicated under the first and third of these diagrams.

Considering each directional diagram as a product of rate constants, the final step is the following: each N_i^∞ is proportional to the sum of the directional diagrams belonging to state i. But, since $\sum_i N_i^\infty = N$,

$$N_i^\infty = Np_i^\infty = \frac{N \times \text{sum of directional diagrams of state } i}{\text{sum of directional diagrams of } \textit{all} \text{ states } (\equiv \Sigma)}. \tag{1.5}$$

In the above example, there are $5 \times 11 = 55$ directional diagrams altogether. Therefore, for state 2, say,

$$p_2^\infty = \frac{N_2^\infty}{N} = \frac{(\alpha_{45}\alpha_{53}\alpha_{31}\alpha_{12} + \alpha_{13}\alpha_{35}\alpha_{54}\alpha_{42} + 9 \text{ other terms})}{(\alpha_{45}\alpha_{53}\alpha_{31}\alpha_{12} + \alpha_{13}\alpha_{35}\alpha_{54}\alpha_{42} + 53 \text{ other terms})}. \tag{1.6}$$

The result 1.5 is intuitively reasonable because the steady-state occupation of the ith state N_i^∞ is proportional to the sum of products of rate constants along different routes leading *toward* state i. That is, we would expect in a general way that the larger the rate constants leading *toward* state i the larger the relative population of state i at $t = \infty$.

Equation 1.5 is, of course, also the solution for N_i^e in the special case that the rate constants correspond to equilibrium at $t = \infty$.

1.4 Simple Examples of the Use of Directional Diagrams

Because there are 55 directional diagrams in the above example, too much space would be required to write out the p_i^∞, as in Eq. 1.6, explicitly. We therefore turn to a few much simpler examples.

Figure 1.2a is the simplest possible steady-state case: a two-state cycle (34). The rate constant notation is indicated in Fig. 1.6a, while Fig. 1.6b shows the two partial diagrams and Fig. 1.6c the four directional diagrams. Following the recipe in Eq. 1.5, we have then

$$p_1^\infty = (\alpha_{21} + \beta_{21})/\Sigma \tag{1.7a}$$

and

$$p_2^\infty = (\alpha_{12} + \beta_{12})/\Sigma, \tag{1.7b}$$

where

$$\Sigma = \alpha_{21} + \beta_{21} + \alpha_{12} + \beta_{12}. \tag{1.8}$$

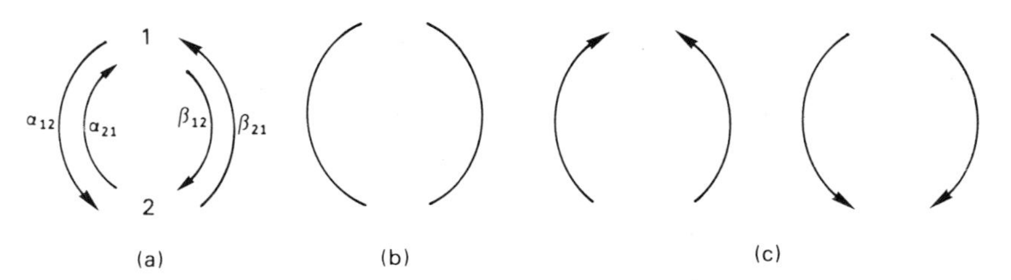

FIG. 1.6 (a) Rate constants for two-state model. (b) Two partial diagrams. (c) Four directional diagrams.

We have equilibrium at $t = \infty$ if the rate constants happen to satisfy $\Pi_+ = \Pi_-$; i.e., if $\alpha_{12}\beta_{21} = \alpha_{21}\beta_{12}$. In this case,

$$\frac{p_1^e}{p_2^e} = \frac{\alpha_{21} + \beta_{21}}{\alpha_{12} + \beta_{12}} = \frac{\alpha_{21}}{\alpha_{12}} = \frac{\beta_{21}}{\beta_{12}} \tag{1.9}$$

(detailed balance).

For the diagram in Fig. 1.1b, we have the three partial diagrams in Fig. 1.7. The arrows (compare Fig. 1.5) can be introduced in three ways, leading to nine directional diagrams. Thus we find, from Eq. 1.5,

$$p_1^\infty = (\alpha_{32}\alpha_{21} + \alpha_{23}\alpha_{31} + \alpha_{21}\alpha_{31})/\Sigma \tag{1.10}$$

$$p_2^\infty = (\alpha_{12}\alpha_{32} + \alpha_{13}\alpha_{32} + \alpha_{31}\alpha_{12})/\Sigma \tag{1.11}$$

$$p_3^\infty = (\alpha_{12}\alpha_{23} + \alpha_{13}\alpha_{23} + \alpha_{21}\alpha_{13})/\Sigma, \tag{1.12}$$

where Σ is the sum of the three numerators (nine terms), that is, the sum of directional diagrams. The condition for equilibrium is $\Pi_+ = \Pi_-$, or $\alpha_{12}\alpha_{23}\alpha_{31} = \alpha_{21}\alpha_{32}\alpha_{13}$.

Figures 1.2d and 1.8 represent a slightly more complicated case. It is easy to see, on introducing arrows into Fig. 1.8 in the four possible ways, that the numerators in p_1^∞, p_2^∞, and p_3^∞ are the same as in Eqs. 1.10–1.12 except that all are multiplied by α_{41}. Also, the numerator for p_4^∞ is as given for p_1^∞ in Eq. 1.10 but multiplied by α_{14}. The new Σ is the sum of the four numerators (12 terms). Thus, the *relative* values of p_1^∞, p_2^∞, and p_3^∞ are the same as in the preceding example (i.e., state 4 does not perturb this steady-state property). Also,

$$p_4^\infty/p_1^\infty = \alpha_{14}/\alpha_{41} = p_4^e/p_1^e. \tag{1.13}$$

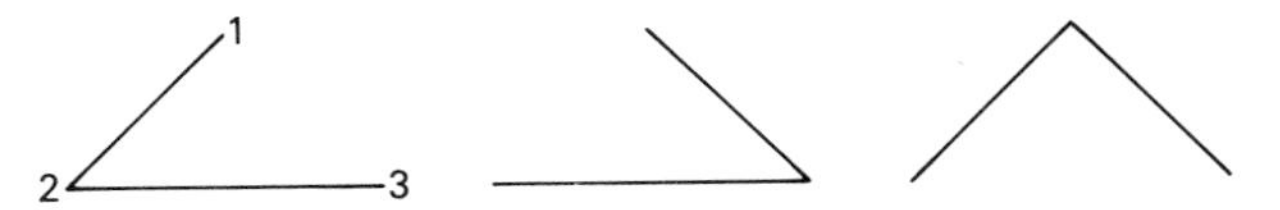

FIG. 1.7 Three partial diagrams for Fig. 1.1b.

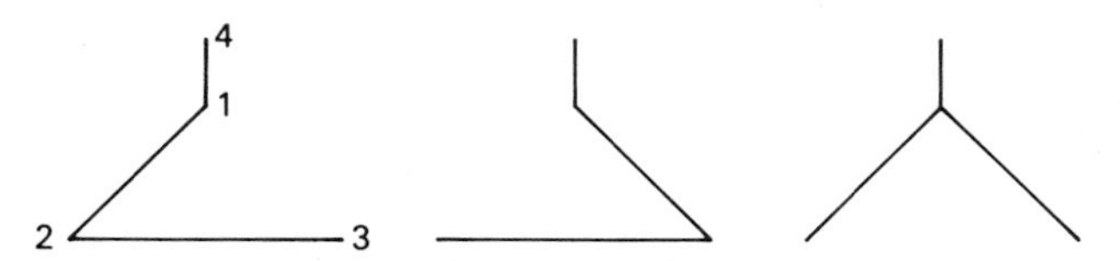

FIG. 1.8 Three partial diagrams for Fig. 1.2d.

That is, states 1 and 4 are in equilibrium with each other even at an arbitrary steady state. This is obvious from $dN_4/dt = 0$ (detailed balance in this case). The condition for complete equilibrium is $\Pi_+ = \Pi_-$, or

$$\alpha_{12}\alpha_{23}\alpha_{31} = \alpha_{21}\alpha_{32}\alpha_{13}$$

as before (irrespective of α_{14} and α_{41}).

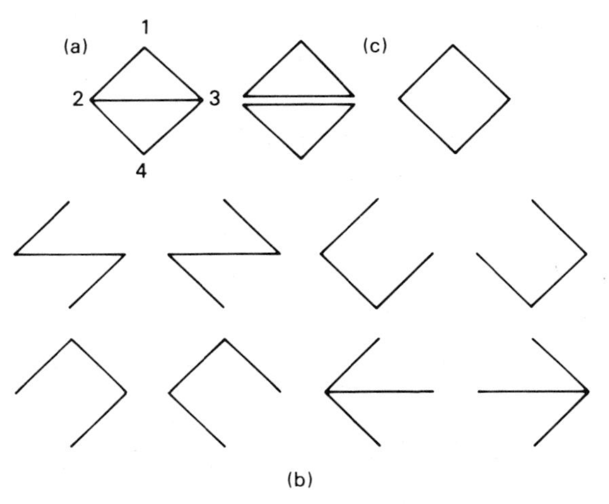

FIG. 1.9 (a) Diagram (example). (b) Eight partial diagrams. (c) Three cycles.

As another example, we merely present, in Fig. 1.9b, the eight partial diagrams belonging to the diagram (32, 35) in Fig. 1.9a. This diagram has three cycles (Fig. 1.9c) and Σ has $4 \times 8 = 32$ terms.

As a final example (36), we consider a simple cycle (Fig. 1.10a) in which one of the rate constants is essentially zero, say, $\alpha_{41} = 0$. Then all directional diagrams that contain this arrow $(1 \leftarrow 4)$ are to be omitted. The four partial diagrams are shown in Fig. 1.10b. The result is simple and intuitively obvious:

$$p_1^\infty \Sigma = [\text{diagram}], \quad p_2^\infty \Sigma = [\text{diagram}] + [\text{diagram}] \tag{1.14}$$

$$p_3^\infty \Sigma = [\text{diagram}] + [\text{diagram}] + [\text{diagram}], \quad p_4^\infty \Sigma = [\text{diagram}] + [\text{diagram}] + [\text{diagram}] + [\text{diagram}],$$

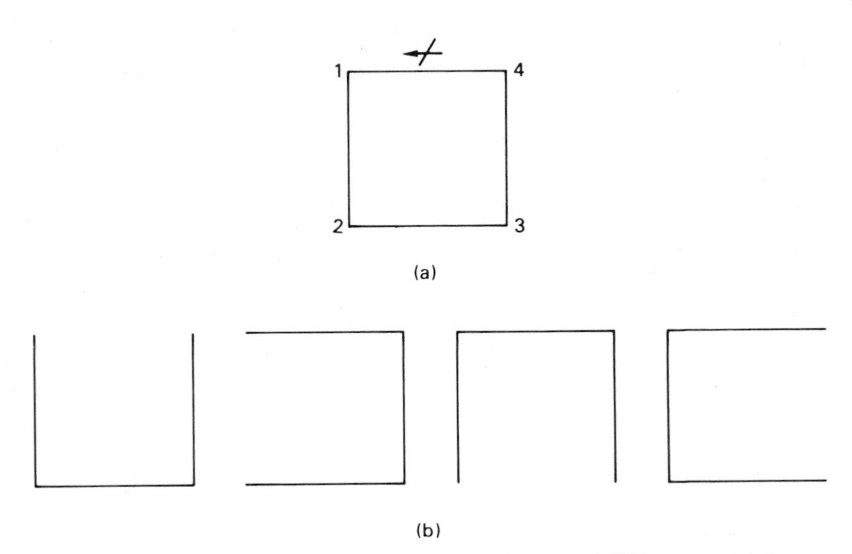

FIG. 1.10 (a) Diagram for four-state single cycle, with $\alpha_{41} = 0$. (b) Four partial diagrams.

where Σ is the sum of the 10 surviving directional diagrams. In the special case in which the seven nonzero rate constants are all equal, the relative populations of the states, in the order 1, 2, 3, 4, would obviously be, from Eqs. 1.14, 1 : 2 : 3 : 4. State 4 has lost none of its directional diagrams while state 1 has lost three out of four.

1.5 Transition Fluxes and Probabilities of States

Let us use Fig. 1.9a for illustrative purposes here. The net mean "transition flux" between states i and j, in the direction $i \to j$, in any diagram is defined as

$$J_{ij} = \alpha_{ij} N_i - \alpha_{ji} N_j = N(\alpha_{ij} p_i - \alpha_{ji} p_j). \tag{1.15}$$

This is the net mean number of transitions of type $i \to j$ per unit time in the ensemble of N units. Note that $J_{ij} \equiv -J_{ji}$. Applying this definition to Fig. 1.9a, we can write the kinetic equations of the system in the form

$$\frac{dN_1}{dt} = J_{21} + J_{31}, \qquad \frac{dN_2}{dt} = J_{12} + J_{32} + J_{42}, \qquad \text{etc.} \tag{1.16}$$

That is, dN_i/dt for each state i is the sum of the J_{ji}'s leading into the state. At equilibrium, each separate $J^e_{ij} = 0$ (detailed balance). At steady state, for Fig. 1.9a,

$$\begin{aligned} &1: \quad J^\infty_{21} + J^\infty_{31} = 0 && 2: \quad J^\infty_{12} + J^\infty_{32} + J^\infty_{42} = 0 \\ &3: \quad J^\infty_{13} + J^\infty_{23} + J^\infty_{43} = 0 && 4: \quad J^\infty_{24} + J^\infty_{34} = 0. \end{aligned} \tag{1.17}$$

Keeping in mind $J_{ij} \equiv -J_{ji}$, one can see that only two of these J's are independent; for example, J^∞_{31} and J^∞_{24}. From these two, all the others can be calculated via Eqs. 1.17.

In the case of a diagram in which all n states are connected in a single cycle (e.g., Fig. 1.1b), we find easily by the same procedure that

$$J^\infty_{12} = J^\infty_{23} = J^\infty_{34} = \cdots = J^\infty_{n1} \,. \tag{1.18}$$

Thus, only one of these J's is independent. This equality of steady-state transition fluxes around a single cycle carries over partially to cases like Figs. 1.3b and 1.9a. For example, for Fig. 1.9a (Eqs. 1.17), $J^\infty_{31} = J^\infty_{12}$ and $J^\infty_{24} = J^\infty_{43}$. That is, in regions where there is no branching in the diagram, steady-state transition fluxes between consecutive pairs of states are equal.

Since each of the n states provides a relation between transition J's (Eq. 1.17), the number of *independent*, nonzero, steady-state transition fluxes for a diagram is $f =$ (*number of lines in diagram*) $- (n - 1)$. For a diagram with a single cycle, $f = 1$.

Transition Fluxes from State Probabilities

If the p^∞_i have been found for any diagram from the directional diagrams, all the transition fluxes J^∞_{ij} can be calculated immediately by means of Eq. 1.15. These are the operational or phenomenological fluxes (Chapter 3): some of them correspond to measurable quantities (e.g., the net rate of transport of a ligand from bath A to bath B across a membrane, as in Fig. 1.3). But from an interpretive or theoretical point of view, these fluxes are incapable, in general, of telling the whole story of the steady cyclic activity taking place within the diagram. For this, individual cycle fluxes (18, 19) are needed—and these are introduced in Chapter 2. In other words, directional diagrams constitute part, but by no means all, of the "diagram method" being summarized in this book. It is the cycle fluxes that are fundamental in irreversible thermodynamics, coupling, fluctuations, etc.

To anticipate Chapter 2 and to verify the above kind of inadequacy of the transition fluxes in a simple, particular case, let us consider Fig. 1.9a. This diagram has five lines and four states. Hence there are two independent transition fluxes calculable from the p_i^∞ (see the discussion of Eq. 1.17): $f = 2$. But this diagram has three cycles and hence three independent cycle fluxes (Chapter 2). One cannot deduce three quantities from two. Of course if the diagram consists of only a single cycle (Figs. 1.1b and 1.10a), there is no problem: there is one independent transition flux and one cycle flux (and they are equal—see Chapter 2).

We conclude with two simple examples (diagram equals a single cycle) of transition fluxes. More complicated cases are handled more easily and elegantly in terms of cycle fluxes; this is reserved for Chapters 2 and 3.

For the two-state cycle in Fig. 1.6a, Eqs. 1.7 and 1.8 give

$$\begin{aligned} J_{12(\alpha)}^\infty &= N(\alpha_{12} p_1^\infty - \alpha_{21} p_2^\infty) \\ &= N(\alpha_{12}\beta_{21} - \alpha_{21}\beta_{12})/\Sigma \\ &= N(\Pi_+ - \Pi_-)/\Sigma = J_{21(\beta)}^\infty . \end{aligned} \tag{1.19}$$

The flux is zero at equilibrium ($\Pi_+ = \Pi_-$), but otherwise it is proportional to $\Pi_+ - \Pi_-$.

Similarly, for the three-state cycle in Fig. 1.1, using Eqs. 1.10–1.12, we find

$$\begin{aligned} J_{12}^\infty &= N(\alpha_{12} p_1^\infty - \alpha_{21} p_2^\infty) \\ &= N(\alpha_{12}\alpha_{23}\alpha_{31} - \alpha_{21}\alpha_{32}\alpha_{13})/\Sigma \\ &= N(\Pi_+ - \Pi_-)/\Sigma = J_{23}^\infty = J_{31}^\infty . \end{aligned} \tag{1.20}$$

Single-Cycle and Multicycle Diagrams

We have begun, in this chapter, the development of a formalism applicable to diagrams of arbitrary complexity. It will already be evident to the reader that single-cycle cases are relatively simple and hardly require such an elaborate theoretical framework (see Appendix 4).

Many proposed models of free energy transduction are presently formulated as single cycles. This simplicity is likely to be temporary in many cases. As can be seen from various examples in this book, multicycle diagrams tend to arise (among other ways) when enzymatic or other macromolecular activity is modified by bound ligands, by conformational changes in the enzyme or macromolecule, or by interactions with other enzymes of a complex (Chapter 7). These effects are very widespread.

As will be obvious from the next two chapters, multicycle diagrams provide an obvious explanation of nonintegral and variable (if substrate concentrations change) operational flux stoichiometry.

REFERENCES

1. D. Oesterhelt and W. Stoeckenius, *Proc. Nat. Acad. Sci. U.S.* **70**, 2853 (1973).
2. R. Henderson and P. N. T. Unwin, *Nature* (*London*) **257**, 28 (1975).
3. J. Keizer, *J. Chem. Phys.* **64**, 1679 (1976).
4. P. D. Boyer, *in* "Dynamics of Energy-Transducing Membranes" (L. Ernster *et al.*, eds.), p. 289. Elsevier, Amsterdam, 1974.
5. P. Mitchell, *J. Bioenerg.* **3**, 5 (1973); **4**, 63 (1974).
6. H. Gutfreund and D. R. Trentham, *in* "Energy Transformation in Biological Systems" (Ciba Symp. 31), p. 69. Elsevier, Amsterdam, 1975.
7. J. C. Skou, *J. Bioenerg.* **4**, 1 (1974).
8. B. Chance, *Ann. N.Y. Acad. Sci.* **227**, 613 (1974).
9. E. C. Slater, *in* "Dynamics of Energy-Transducing Membranes" (L. Ernster *et al.*, eds.), p. 1. Elsevier, Amsterdam, 1974.
10. E. Racker, *in* "Dynamics of Energy-Transducing Membranes" (L. Ernster *et al.*, eds.), p. 269. Elsevier, Amsterdam, 1974.
11. V. P. Skulachev, *Ann. N.Y. Acad. Sci.* **227**, 188 (1974).
12. M. Klingenberg, *in* "Dynamics of Energy-Transducing Membranes" (L. Ernster *et al.*, eds.), p. 511. Elsevier, Amsterdam, 1974.
13. T. G. Ebrey and B. Honig, *Q. Rev. Biophys.* **8**, 129 (1975).
14. H. T. Witt, *Q. Rev. Biophys.* **4**, 365 (1971).
15. T. L. Hill, *Progr. Biophys. Mol. Biol.* **28**, 267 (1974).
16. T. L. Hill, *Progr. Biophys. Mol. Biol.* **29**, 105 (1975).
17. E. L. King and C. Altman, *J. Phys. Chem.* **60**, 1375 (1956).
18. T. L. Hill, *J. Theoret. Biol.* **10**, 442 (1966).
19. T. L. Hill, "Thermodynamics for Chemists and Biologists," Chapter 7. Addison-Wesley, Reading, Massachusetts, 1968.
20. T. L. Hill, *Biochemistry* **14**, 2127 (1975).
21. T. L. Hill and R. M. Simmons, *Proc. Nat. Acad. Sci. U.S.* **73**, 95 (1976).
22. T. L. Hill and R. M. Simmons, *Proc. Nat. Acad. Sci. U.S.* **73**, 336, 2165 (1976).
23. R. M. Simmons and T. L. Hill, *Nature* (*London*) **263**, 615 (1976).
24. T. L. Hill, *J. Chem. Phys.* **54**, 34 (1971).
25. T. L. Hill and Y. Chen, *Proc. Nat. Acad. Sci. U.S.* **72**, 1291 (1975).
26. Y. Chen and T. L. Hill, *Biophys. J.* **13**, 1276 (1973).
27. Y. Chen, *J. Chem. Phys.* **59**, 5810 (1973).
28. Y. Chen, *J. Theoret. Biol.* **55**, 229 (1975).
29. Y. Chen, *Proc. Nat. Acad. Sci. U.S.* **72**, 3807 (1975).
30. T. L. Hill, *Proc. Nat. Acad. Sci. U.S.* **73**, 4432 (1976); T. L. Hill, *in* "Statistical Mechanics and Statistical Methods in Theory and Application" (U. Landman, ed.). Plenum, New York, 1977.
31. Y. Chen, *Adv. Chem. Phys.* (in press).
32. T. L. Hill and O. Kedem, *J. Theoret. Biol.* **10**, 399 (1966).
33. T. L. Hill and Y. Chen, *Proc. Nat. Acad. Sci. U.S.* **68**, 1711 (1971).
34. T. L. Hill, E. Eisenberg, Y. Chen, and R. J. Podolsky, *Biophys. J.* **15**, 335 (1975).
35. T. L. Hill, *Proc. Nat. Acad. Sci. U.S.* **55**, 1379 (1966).
36. T. L. Hill, *Proc. Nat. Acad. Sci. U.S.* **65**, 409 (1970).

Chapter 2

The Diagram Method: Cycles

2.1 Cycles and Cycle Fluxes

Here we extend the methodology of Chapter 1 by the explicit introduction of the cycles and cycle fluxes that belong to a diagram (1, 2). Cycle fluxes are the fundamental *theoretical* components of the (operational) transition fluxes used in Chapter 1. We resort again to the model in Figs. 1.3 and 2.1a to introduce this subject. This model has three cycles, shown in Fig. 2.1b. Note that cycle *a* transports both L_1 and L_2 between baths A and B, cycle *b* transports L_1, and cycle *c* transports L_2.

A *cyclic diagram* is obtained from a directional diagram by adding one arrow (i.e., multiplying by a rate constant) in a direction such as to complete a closed path of arrows, a cycle, all in the same direction. Cyclic diagrams occur in pairs, as shown in Fig. 2.2. These differ only in the direction of traversal of the cycle. The algebraic difference between such a pair of cyclic diagrams is a *flux diagram.* The difference may be taken in two ways, of course. All possible flux diagrams for this model are shown in Fig. 2.3. For example, from Fig. 2.2, the algebraic value of flux diagram II in Fig. 2.3 is (if + signifies counterclockwise)

$$\mathrm{II} = (\alpha_{12}\alpha_{23}\alpha_{31} - \alpha_{21}\alpha_{32}\alpha_{13})\alpha_{45}\alpha_{53}\,. \tag{2.1}$$

If i and j are adjacent states in a diagram, the transition flux $j \to i$ is

$$J_{ji}^{\infty} = \alpha_{ji}N_j^{\infty} - \alpha_{ij}N_i^{\infty}.$$

Then, in view of the relation between directional and flux diagrams, we have (as proved in Appendix 2)

$$J_{ji}^{\infty} = \frac{N \times \text{sum of } j \to i \text{ flux diagrams}}{\text{sum of directional diagrams of all states } (\equiv \Sigma)} \tag{2.2}$$

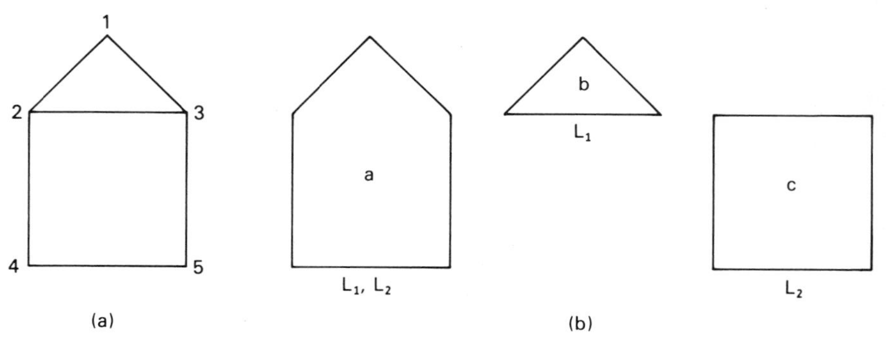

FIG. 2.1 (a) Diagram from Fig. 1.3. (b) Three cycles, with transported ligands indicated.

where a flux diagram is considered to be a $j \to i$ flux diagram if the line i–j is included in the cycle of the flux diagram. Explicitly, to establish the algebraic sign,

$$(j \to i \text{ flux diagram}) = (j \to i \text{ cyclic diagram}) - (i \to j \text{ cyclic diagram}). \tag{2.3}$$

For example, in computing J_{12}^{α} for our present model, the $1 \to 2$ flux diagrams are I, II, III, and IV in Fig. 2.3. Incidentally, these can be generated by adding the line 1–2 to those partial diagrams in Fig. 1.4 with the 1–2 line missing (also labeled I to IV). The sign chosen in Eq. 2.1 for flux diagram II is correct for J_{12}^{α}.

A flux diagram may be derived from a partial diagram by adding one line, as just mentioned. Alternatively, we can define a flux diagram as a diagram that contains one and only one cycle (with arrows omitted since the cycle is traversed in both directions) plus one or more streams (with arrows included) flowing into the cycle at one or more points of the cycle. Streams may converge, but may not diverge on their way toward the cycle.

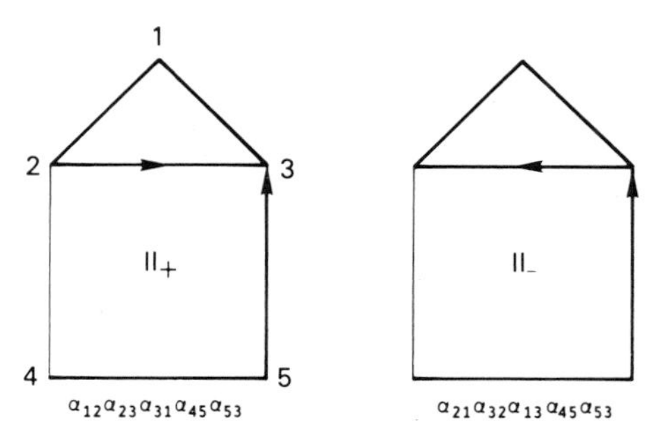

FIG. 2.2 A pair of cyclic diagrams, with algebraic values. See Figs. 1.4 and 2.3.

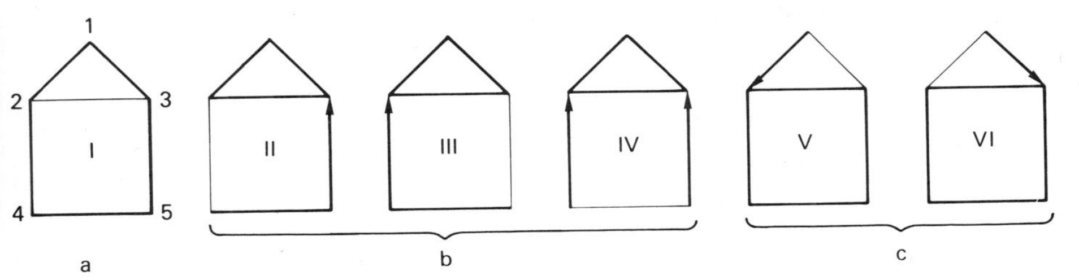

FIG. 2.3 Six flux diagrams, classified according to cycle.

The number of lines used in a flux diagram is one more than the number of lines used in a partial diagram. A flux diagram is related to a cycle in the same way that a directional diagram is related to a state. The flux diagram sum in Eq. 2.2 is the sum over all those diagrams of the type just described whose cycle includes the line i–j. Each flux diagram makes an additive contribution to the total flux in Eq. 2.2.

Nonvanishing steady-state fluxes are possible only for systems whose diagrams contain one or more cycles. Alteration of an equilibrium set of rate constants in a diagram with no cycles can lead only to a new equilibrium state at $t = \infty$. If the diagram has one or more cycles, the rate constant alteration from the equilibrium set must occur in one of the cycles in order to produce a nonequilibrium steady state.

Different flux diagrams that contain the *same* cycle may be collected and added together. The sum, for brevity, is also referred to as a *cycle* (always in an algebraic context). Thus, noting the labeling in Figs. 2.1b and 2.3, we have

$$\text{cycle } a = \text{I}, \qquad \text{cycle } b = \text{II} + \text{III} + \text{IV}, \qquad \text{cycle } c = \text{V} + \text{VI}. \tag{2.4}$$

When adding several flux diagrams, the same directionality must be used in all terms (usually + signifies counterclockwise). The *cycle fluxes* J_κ ($\kappa = a, b, c$) are then defined by

$$J_a = N(\text{cycle } a)/\Sigma, \qquad J_b = N(\text{cycle } b)/\Sigma, \tag{2.5}$$

etc. Since cycle fluxes are defined only at steady state, we can omit the usual superscript. Cycle fluxes can have positive or negative values.

Returning now to Eq. 2.2, we see that we can also write

$$J_{ji}^{\infty} = N \times (\text{sum of } j \to i \text{ cycles})/\Sigma, \tag{2.6}$$

where the sum in the numerator is over all cycles that include the line i–j as part of the cycle. The sign of a cycle in Eq. 2.6 is positive if its assigned + direction agrees with $j \to i$; otherwise it is negative. In view of Eqs. 2.5, we have then

$$J_{ji}^{\infty} = \text{sum of } j \to i \text{ cycle fluxes}, \tag{2.7}$$

with the same rule for signs as in Eq. 2.6. Thus, in our example (+ is taken as counterclockwise for each cycle),

$$J_{12}^{\infty} = J_a + J_b, \qquad J_{23}^{\infty} = J_b - J_c, \qquad J_{24}^{\infty} = J_a + J_c. \tag{2.8}$$

Also,

$$J_{31}^{\infty} = J_{12}^{\infty}, \qquad J_{24}^{\infty} = J_{45}^{\infty} = J_{53}^{\infty}. \tag{2.9}$$

Figure 1.3a provides the obvious physical significance of Eqs. 2.9: the first flux gives the rate of gain of L_1 by bath B, or the rate of loss of L_1 by bath A; and the second flux gives the rate of loss of L_2 by bath A, or the rate of L_2 crossing the membrane; or the rate of gain of L_2 by bath B.

We note that, as mentioned in Section 1.5, there are three independent cycle fluxes here but only two independent transition fluxes. Cycle fluxes represent a more detailed subdivision or breakdown of a diagram's steady-state kinetic activity than do the (operational) transition fluxes. Cycle fluxes result from a theoretical construct: the classification of flux diagrams according to cycle type. But this construct makes possible, as we shall see, a deeper understanding of the steady-state kinetics of these systems. As an immediate example: the fact that J_{12}^{∞} (rate of disappearance of L_1 from bath A) and J_{24}^{∞} (same for L_2) both contain the component J_a (Eqs. 2.8), which involves transport of *both* L_1 and L_2 (Fig. 2.1b), guarantees that these two operational fluxes are *coupled* in the thermodynamic sense (Chapter 3) for this model. In other words, active transport and free energy transduction are possible. Indeed, this follows from the simple fact that the diagram for this model contains at least one cycle (cycle *a*, Fig. 2.1b) that transports both species; no equations are required in order to see this.

Although Eq. 2.7 has been arrived at to some extent in the context of a special case, it obviously applies to any diagram since Eqs. 2.4–2.7 simply concern *classification* of flux diagrams. *To summarize*: any transition flux can be written as a sum of contributions (cycle fluxes) from each cycle in the diagram that includes that transition as part of the cycle; and, in turn, each cycle flux can be written as N/Σ multiplied by the sum of those flux diagrams that belong to the given cycle.

The cycle fluxes and flux diagrams just mentioned are properties of the diagram as a whole; they do not "belong" to any particular transition *ji*. More specifically, suppose a cycle includes the successive states *j*–*i*–*l*. The contribution of the cycle (namely, the cycle flux) to J_{ji}^{∞} is the same quantity as the contribution of the cycle to J_{il}^{∞}, since both of these contributions are proportional to the sum of the same set of flux diagrams. Therefore a cycle makes the *same* contribution to the transition flux (in the same direction) along *all* lines making up the cycle. This is a rather obvious but important conclusion (Chapter 3).

The sum of flux diagrams belonging to a given cycle determines the relative magnitude of this cycle flux (compared to the other cycle fluxes of the diagram). This is an intuitively reasonable result in essentially the same way that Eq. 1.5 is. The flux associated with a cycle would be expected to be larger the greater the difference in the product of the rate constants going around the cycle in the two different directions (Π_+ and Π_-), and the larger the rate constants leading *toward* ("feeding into") the cycle from other parts of the diagram (see Fig. 2.3 and Eq. 2.1).

The relative steady-state populations of states are determined by Eq. 1.5; the relative steady-state cycle fluxes are determined in an analogous fashion by Eqs. 2.4 and 2.5. As we shall see, the latter quantities (J_κ) have more interpretive value than the former (p_i^∞). Incidentally, if only relative values of the populations and cycle fluxes are needed, the usually lengthy denominator Σ in Eqs. 1.5 and 2.5 can be omitted.

Let us return now to our example and to Eqs. 2.4 and 2.5 in order to make the algebra more explicit. We find, from Fig. 2.3 (+ signifies counterclockwise),

$$\begin{aligned}
\text{cycle } a &= \text{I} = \Pi_{a+} - \Pi_{a-} \\
\text{cycle } b &= \text{II} + \text{III} + \text{IV} \\
&= (\Pi_{b+} - \Pi_{b-})(\alpha_{45}\alpha_{53} + \alpha_{54}\alpha_{42} + \alpha_{42}\alpha_{53}) \\
\text{cycle } c &= \text{V} + \text{VI} = (\Pi_{c+} - \Pi_{c-})(\alpha_{12} + \alpha_{13}),
\end{aligned} \tag{2.10}$$

where $\Pi_{a+} = \alpha_{12}\alpha_{24}\alpha_{45}\alpha_{53}\alpha_{31}$, $\Pi_{b-} = \alpha_{21}\alpha_{32}\alpha_{13}$, etc. Generalizing, we see that, for any diagram and for any cycle κ belonging to that diagram, the cycle flux J_κ can always be written in the form

$$J_\kappa = N(\Pi_{\kappa+} - \Pi_{\kappa-})\Sigma_\kappa/\Sigma, \tag{2.11}$$

where Σ_κ is a sum, over all flux diagrams belonging to κ, of those parts of the flux diagrams that *feed into* the cycle (i.e., these are appendages to the cycle). Thus, in Fig. 2.3,

$$\Sigma_a = 1, \qquad \Sigma_b = \alpha_{45}\alpha_{53} + \alpha_{54}\alpha_{42} + \alpha_{42}\alpha_{53}, \qquad \Sigma_c = \alpha_{12} + \alpha_{13}. \tag{2.12}$$

In Eq. 2.11, $(\Pi_{\kappa+} - \Pi_{\kappa-})\Sigma_\kappa$ is the sum of flux diagrams for cycle κ, $\Pi_{\kappa+}\Sigma_\kappa$ is the sum of + cyclic diagrams for cycle κ, and J_κ/N is a function of rate constants only—but of *all* the rate constants in the diagram (because of Σ).

2.2 One-Way Cycle Fluxes; Kinetics at the Cycle Level

If the argument in Appendix 2, leading to Eq. 2.2, is reviewed, one concludes that the *one-way* transition flux $\alpha_{ji} N_j^{\infty}$ between states i and j in the direction $j \rightarrow i$ is given by $N \times$ (sum of $j \rightarrow i$ *cyclic* diagrams)/Σ (see, e.g., Fig. 2.2). To see this, it is necessary only to split *pairs* of cyclic diagrams into *separate* cyclic diagrams. If we then repeat the argument subsequent to Eq. 2.2, replacing flux diagrams by cyclic diagrams, we obtain for a final result, corresponding to Eq. 2.11, that

$$J_{\kappa+} = N\Pi_{\kappa+}\Sigma_{\kappa}/\Sigma \qquad \text{and} \qquad J_{\kappa-} = N\Pi_{\kappa-}\Sigma_{\kappa}/\Sigma \tag{2.13}$$

are the *separate one-way* cycle fluxes. Also, $J_{\kappa} = J_{\kappa+} - J_{\kappa-}$. Both $J_{\kappa+}$ and $J_{\kappa-}$ are positive quantities; J_{κ} can be positive or negative. $J_{\kappa+}$ is the number of cycles of type κ completed in the + direction per unit time, in the ensemble of N systems. The exact meaning of "completed" will be dealt with in Chapter 6. This result (Eq. 2.13) is not self-evident in advance since a difference $x - y$ might have separate components $x + c$ and $y + c$, where $c \neq 0$. The result 2.13 is confirmed by noting that if, in Eq. 2.11, any one of the rate constants in the product $\Pi_{\kappa-}$ is set equal to zero (so that κ cycles can be completed in the + direction only), we then obtain $J'_{\kappa+} = N\Pi_{\kappa+}\Sigma_{\kappa}/\Sigma'$, where the primes indicate that those terms in Σ which contain the particular rate constant set equal to zero will be missing. Equations 2.13 have also been checked by Monte Carlo computer calculations in special cases (3) (Chapter 6).

One-way transition fluxes can be observed in some cases by means of radioactive tracers. One-way cycle fluxes are even less operational than cycle fluxes, insofar as mean values are concerned. But the one-way cycle fluxes *are* involved in observable fluctuations and noise in transition fluxes (Chapter 6).

Rate Constants and Kinetics at Two Levels of Detail

As will be made more explicit below (and especially in Chapter 6), the result 2.13 suggests that the kinetic events of a steady-state diagram can be viewed at two different levels of detail (4): the individual transition level (rate constants α_{ij}) or *the complete cycle level* (cycle rate constants $k_{\kappa+} \equiv J_{\kappa+}/N$, $k_{\kappa-} \equiv J_{\kappa-}/N$). Actually, in the theoretical treatment of muscle contraction (5) (Chapter 5), one is generally obliged to follow individual transitions for two reasons: an analysis of the conversion of chemical free energy into mechanical work requires attention to single transitions; and the rate constants of the diagram are functions of a spatial variable x

which itself varies linearly with time in steady contractions. But in membrane transport and other biochemical problems the transition probabilities α_{ij} are constant (if ligand concentrations are constant) and the required free energy conversion (transduction) bookkeeping is related to completed cycles only and not to individual transitions within cycles (Chapter 6). Thus, in these problems, besides considering individual states and transitions *within* each cycle, as an alternative one can work at a level somewhat less detailed. That is, one can deal with *composite* cycle transition probabilities $k_{\kappa+}$, $k_{\kappa-}$ for completion of each kind of cycle in the diagram, in either direction, together with the overall free energy changes associated with entire cycles. This latter approach is of course one step more closely related to the use of conventional thermodynamic *net* fluxes and forces for the system (Chapter 3).

There may be systems in the study of which one must be content (temporarily) to work, both theoretically and experimentally, entirely in terms of cycle rate constants rather than individual transition rate constants. The simplest example: for a membrane transport system with a single cycle *a*, radioactive tracer studies plus the net flux J_a would provide the two cycle rate constants k_{a+} and k_{a-}.

Cycle Kinetics

We proceed now to introduce additional basic ideas involved in cycle kinetics (4), but we reserve numerical examples and details for Chapters 4 and 6. As already mentioned, we define, for any cycle κ, the first-order cycle rate constants $k_{\kappa+}$ and $k_{\kappa-}$ by

$$\begin{aligned} k_{\kappa+} &= J_{\kappa+}/N = \Pi_{\kappa+}\Sigma_\kappa/\Sigma \\ k_{\kappa-} &= J_{\kappa-}/N = \Pi_{\kappa-}\Sigma_\kappa/\Sigma \\ k_{\kappa+}/k_{\kappa-} &= J_{\kappa+}/J_{\kappa-} = \Pi_{\kappa+}/\Pi_{\kappa-}. \end{aligned} \tag{2.14}$$

Thus the cycle rate constants are composite, explicit combinations of the transition rate constants α_{ij} of the diagram. Recall that Σ, in Eqs. 2.14, involves every α_{ij} in the diagram. But the *ratio* $k_{\kappa+}/k_{\kappa-}$ depends only on the rate constants around the cycle κ itself. As an example of a cycle rate constant (Fig. 2.3), from Eqs. 1.6 and 2.10,

$$k_{b+} = \frac{\alpha_{12}\alpha_{23}\alpha_{31}(\alpha_{45}\alpha_{53} + \alpha_{54}\alpha_{42} + \alpha_{42}\alpha_{53})}{\alpha_{45}\alpha_{53}\alpha_{31}\alpha_{12} + 54 \text{ other terms}}. \tag{2.15}$$

The rate at which a given transition $j \rightarrow i$ occurs in the ensemble of N systems at steady state is $\alpha_{ji} N_j^\infty$; the rate at which κ cycles are completed in the $+$ direction is $k_{\kappa+} N$. Note the N_j^∞ in the former case (only part of the population) and the N in the latter. After a transition, the state of the system has been changed; after a cycle is completed, the system is unaltered (except for the transport, etc., accomplished). This last statement is adequate for present purposes but is examined much more carefully in Chapter 6.

Having introduced the cycle rate constants, it is instructive to take a stochastic view of the steady-state kinetics. Imagine (and this can be simulated on a computer—see Chapter 6) that we follow in detail a single system of the ensemble over a long period of time, as it occasionally and instantaneously changes from one state to another of its diagram, in accordance with the first-order transition probabilities α_{ij} of the diagram. Suppose the diagram has cycles $\kappa = a, b, c, \ldots$. Actually, starting in any arbitrary state, what we record as time passes is the completion of each successive cycle (the type of cycle; its direction, $\pm$; and the time between completed cycles). Over an extremely long period of time, let $p_{a+}, p_{a-}, p_{b+}, p_{b-}, \ldots$ be the fraction of completed cycles of type a in direction $+$, etc. The sum of these probabilities is unity. Also, let τ be the mean time between cycle completions (total time divided by the total number of cycles completed of any kind). Then the probability of completing any cycle in the infinitesimal interval dt is dt/τ, while the probability of completing a cycle of type $a+$, $a-$, etc., in dt is $p_{a+}\ dt/\tau$, etc. The relations between the $k_{\kappa\pm}$ and the $p_{\kappa\pm}$ are therefore

$$k_{\kappa+} = p_{\kappa+}/\tau, \qquad k_{\kappa-} = p_{\kappa-}/\tau, \qquad \text{etc.} \tag{2.16}$$

The cycle probabilities can then be written

$$p_{a+} = \tau k_{a+} = \tau \Pi_{a+} \Sigma_a / \Sigma, \qquad p_{a-} = \tau \Pi_{a-} \Sigma_a / \Sigma, \tag{2.17}$$

etc. Since the sum of these probabilities is unity,

$$\begin{aligned} \tau &= (k_{a+} + k_{a-} + k_{b+} + \cdots)^{-1} \\ &= \Sigma / (\Pi_{a+}\Sigma_a + \Pi_{a-}\Sigma_a + \Pi_{b+}\Sigma_b + \cdots). \end{aligned} \tag{2.18}$$

That is, τ, the mean time between completed cycles, is equal to the sum of all directional diagrams divided by the sum of all cyclic diagrams. The total cycle completion frequency for a single system of the ensemble is τ^{-1}.

The cycle probabilities can also be written

$$p_{a+} = \Pi_{a+}\Sigma_a / (\Pi_{a+}\Sigma_a + \Pi_{a-}\Sigma_a + \Pi_{b+}\Sigma_b + \cdots), \tag{2.19}$$

etc. Thus p_{a+} is the sum of cyclic diagrams belonging to $a+$ divided by the sum of *all* cyclic diagrams. Note the close analogy to Eq. 1.5 for the state probabilities p_i^{∞}.

As an explicit example of Eq. 2.19, consider Fig. 2.3 and Eqs. 2.10:

$$p_{c+} = \frac{\Pi_{c+}(\alpha_{12} + \alpha_{13})}{\Pi_{a+} + \Pi_{a-} + (\Pi_{b+} + \Pi_{b-})(\alpha_{45}\alpha_{53} + \alpha_{54}\alpha_{42} + \alpha_{42}\alpha_{53}) + (\Pi_{c+} + \Pi_{c-})(\alpha_{12} + \alpha_{13})}, \tag{2.20}$$

etc. All terms are products of five rate constants.

This subject is pursued further in Chapters 4 and, especially, 6.

2.3 Further Examples of Cycles and of Flux Diagrams

Suppose, in Figs. 1.3ab and Fig. 2.1a, that there is a relatively fast equilibrium between states 2 and 3 (conformational change in E when L_1 is bound). The diagram then simplifies to that shown in Fig. 2.4. There are now only two cycles: cycle *b* transports L_1; and cycle *c* transports L_2. The coupling that exists in Fig. 2.1a is now absent because cycle *a* is missing; active transport of one ligand by the other is no longer possible [see also Hill (2), p. 164]. This model (Fig. 2.4) has six partial diagrams, $4 \times 6 = 24$ directional diagrams, and five flux diagrams. The number of independent transition fluxes is $f = 5 - 3 = 2$.

As a second example, consider a reduced version of Fig. 1.2c. Recall that, for example, states 1, 2, 3 might be E, ES, EP and states 4, 5, 6 might be LE, LES, LEP. Suppose that the ligand L increases rate constants in its cycle to such an extent that, on the time scale of states 1, 2, 3, there is a fast equilibrium between states 4, 5, 6 (i.e., in effect, these are reduced to a single state).

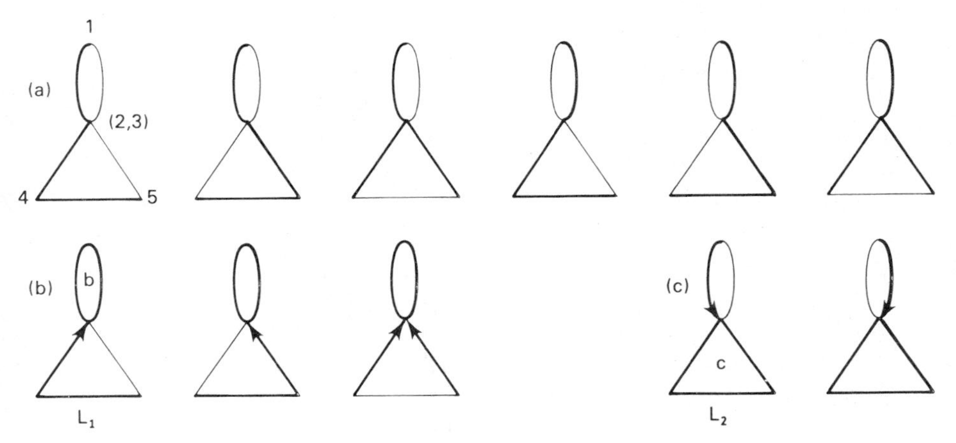

FIG. 2.4 The diagram in Figs. 1.3b and 2.1a simplified by a fast equilibrium between states 2 and 3. (a) Partial diagrams. (b) Flux diagrams for cycle *b*. (c) Flux diagrams for cycle *c*.

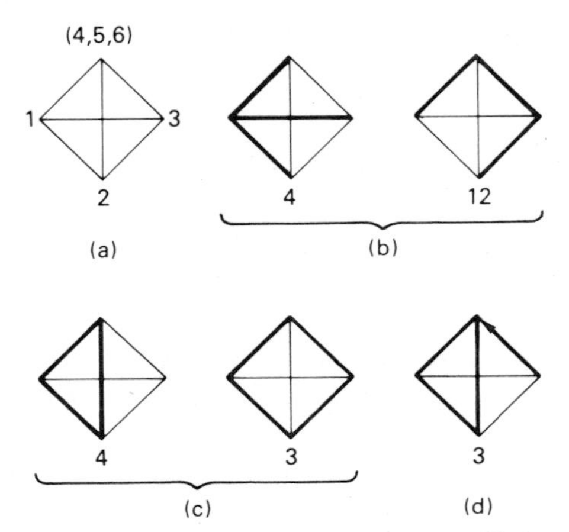

FIG. 2.5 (a) The diagram in Fig. 1.2c reduced by a fast equilibrium between states 4, 5, 6. (b) Sixteen partial diagrams. (c) Seven cycles. (d) Three flux diagrams for the cycle shown.

The diagram is then a tetrahedron (Fig. 2.5a). The diagram has 16 partial diagrams (Fig. 2.5b), $4 \times 16 = 64$ directional diagrams, seven cycles (Fig. 2.5c), and 15 flux diagrams (each of the four triangular cycles has three flux diagrams, as indicated in Fig. 2.5d). Also, $f = 6 - 3 = 3$.

Finally, we consider a more complicated diagram that can be used to illustrate active transport of Na^+ and K^+ across a cell membrane. This model, which is a generalization of Fig. 1.3, will be examined in more detail in the next chapter. Figure 2.6 shows the eight possible states of a unit. The enzyme E (a *hypothetical* Na, K-ATPase) exists in two conformations, has binding sites for Na^+ and K^+, and can be phosphorylated (transition $5 \rightarrow 7$) via binding of ATP, fast hydrolysis of ATP, and fast release of ADP to the inside (i.e., some transient intermediates—see Appendix 1—are omitted). The conformational change is assumed possible in any binding state of E. P in Fig. 2.6 can be released only to the outside (transition $8 \rightarrow 6$). This model is a somewhat more realistic modification of one used in earlier publications (1, 2).

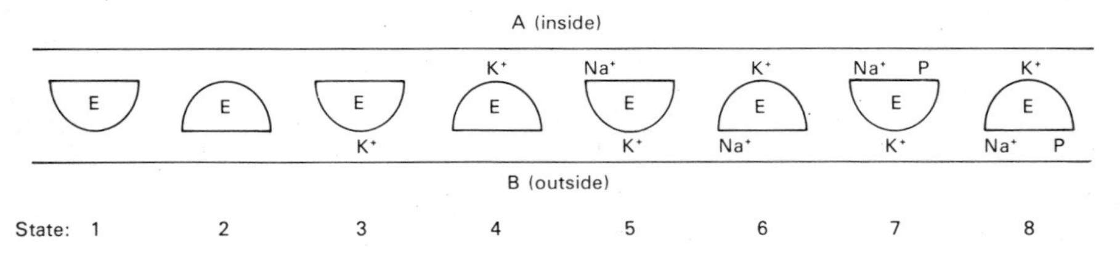

FIG. 2.6 Illustrative model for active transport governed by a Na,K-ATPase. E is ATPase with two conformations; P is bound phosphate; etc.

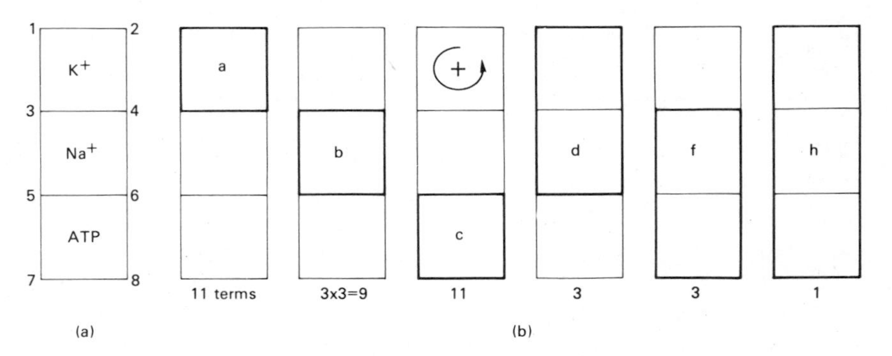

FIG. 2.7 (a) Diagram for Fig. 2.6. (b) Six cycles, with number of flux diagrams per cycle indicated.

The diagram is shown in Fig. 2.7a. There are six cycles (Fig. 2.7b), two of which (*f*, *h*) couple the ATP reaction with Na^+ and K^+ transport. Hence active transport is possible in this model, with ATP as the free energy source. Presumably cycle *h* would be the most important of the six cycles.

From Fig. 2.7 it is apparent that the K^+ flux across the membrane receives contributions from cycles *a*, *d*, and *h*, the Na^+ flux from cycles *b*, *d*, *f*, and *h*, and the ATP flux from cycles *c*, *f*, and *h* (see Chapter 3 for details).

There are 38 flux diagrams, as indicated under the cycles in Fig. 2.7b. These are all obvious except for cycles *a* and *c*, which are formally equivalent. The 11 flux diagrams for cycle *a* are given in Fig. 2.8.

The diagram for this model has a large number of partial diagrams (namely, 56). They are perhaps most easily classified according to which of the four horizontal lines in the diagram are used in a given partial diagram (each partial diagram has seven lines altogether). Figure 2.9 shows the breakdown of the partial diagrams in this way. The number of possible partial diagrams for each class is indicated. For example, on the left of this figure, six vertical lines must be added to complete a partial diagram; on the

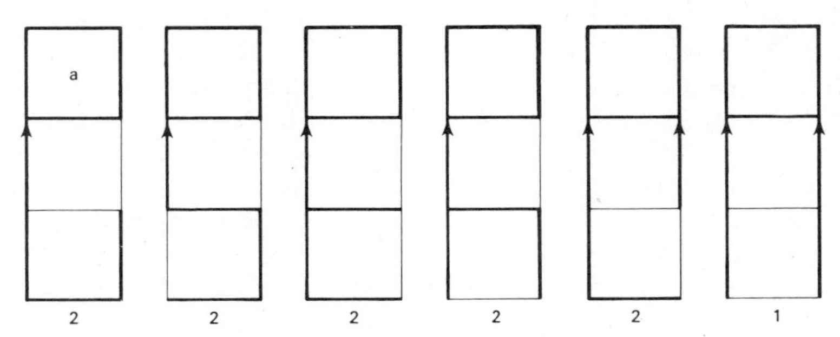

FIG. 2.8 Eleven flux diagrams for cycle *a*.

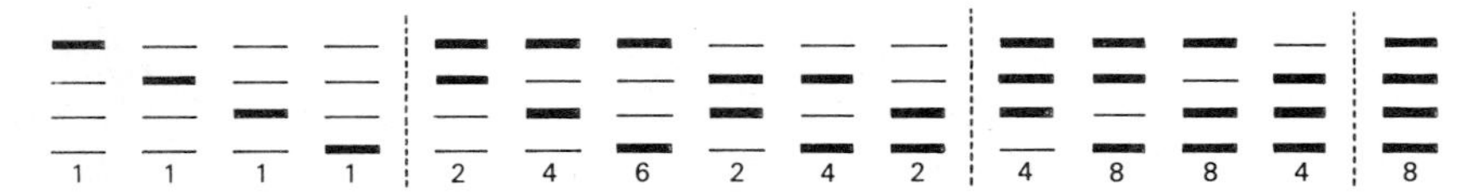

FIG. 2.9 Fifty-six partial diagrams classified according to the horizontal lines (heavy) that appear in the partial diagrams.

right, three must be added. The total number of directional diagrams is $8 \times 56 = 448$.

The number of independent transition fluxes is $f = 10 - 7 = 3$. Thus we are using the maximum possible number of phenomenological fluxes for this model (K^+, Na^+, ATP).

Charged particles and a membrane potential are involved here. In general, all the rate constants of the model might be expected to contain electrostatic contributions (Appendix 3). But these would appear only in a very explicit molecular model. At the formal level being used in this book, we can manipulate the rate constants α_{ij} of a model without enquiring into the molecular theory of the α_{ij} themselves.

2.4 Calculation of State Probabilities from Cycle Fluxes

As we have just seen in the last example, the number of directional diagrams might easily become too large for practical algebraic purposes. In extreme cases of this sort, a possibility (6) is to obtain approximate *algebraic* expressions by a Monte Carlo selection of a reasonable sample of directional diagrams out of the prohibitive total. But we shall not pursue this here. Instead, we discuss a method (4) that is often useful (a) if the number of flux diagrams is modest while the number of directional diagrams is large, and (b) if completely general algebraic results are not needed—i.e., for completely arbitrary α_{ij}. The method can in fact be used if general algebraic results are required, but in this case it may or may not offer an advantage, depending on how difficult it is to enumerate the directional diagrams.

In this method we completely avoid the explicit consideration of directional diagrams. Instead the state probabilities are found by combining the unnormalized cycle fluxes with relations between state probabilities and transition fluxes. The smaller *number* of flux diagrams compared to directional diagrams is not the only advantage here; in addition, the flux diagrams are much easier to write out than directional diagrams since, in effect, the cycle itself is removed from the diagram. That is, in a flux diagram, arrows can be inserted only on those lines of the diagram not already preempted by the cycle in question (compare, e.g., Figs. 1.5 and 2.3).

As a first step, we must obtain the cycle fluxes, except for the factor Σ. We then select a small cycle in the diagram, such as the cycle 1231 in Fig. 2.10. In this cycle, we first express, say, p_2^∞ and p_3^∞ (unnormalized) in terms of the cycle fluxes contributing to J_{23}^∞. We then do the same for p_3^∞, p_1^∞ and p_2^∞, p_1^∞. By eliminating p_1^∞ from the latter two expressions, we find a second connection between the two unknowns p_2^∞ and p_3^∞. We can then solve for the unnormalized p_2^∞ and p_3^∞ separately. With these quantities available, one can then spread the solution over the entire diagram, state by state: for example, we can get p_1^∞ from J_{12}^∞, p_4^∞ from J_{24}^∞, etc. Finally, we can use the normalization of the p_i^∞ to obtain Σ, the normalization constant. This same Σ, of course, completes the solution for the cycle fluxes, J_κ.

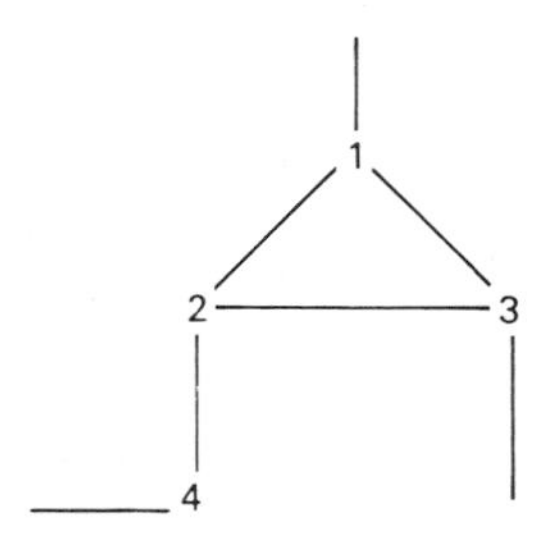

FIG. 2.10 Part of a sample diagram.

To see how the method works in detail, we turn, as usual, to an example (4). This is based on Fig. 1.3c, but we renumber the states and choose $L_1 = Na^+$ and $L_2 = L$. That is, we have in mind a hypothetical model for the active transport of the ligand L (e.g., a sugar or amino acid) by a Na^+ electrochemical gradient (the membrane potential need not be introduced explicitly for present purposes). The case of primary interest occurs when the downhill Na^+ electrochemical gradient (out → in) is large enough to drive the ligand uphill (out → in) against its concentration gradient.

The states and diagram are shown in Fig. 2.11. Figure 2.12 exhibits the

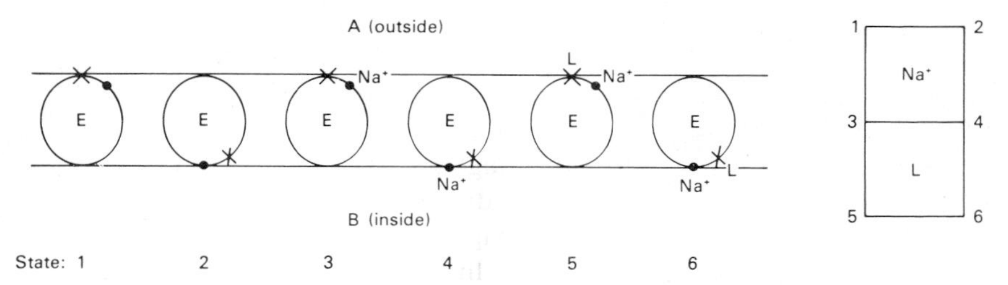

FIG. 2.11 Illustrative model and diagram for active transport of L by Na^+.

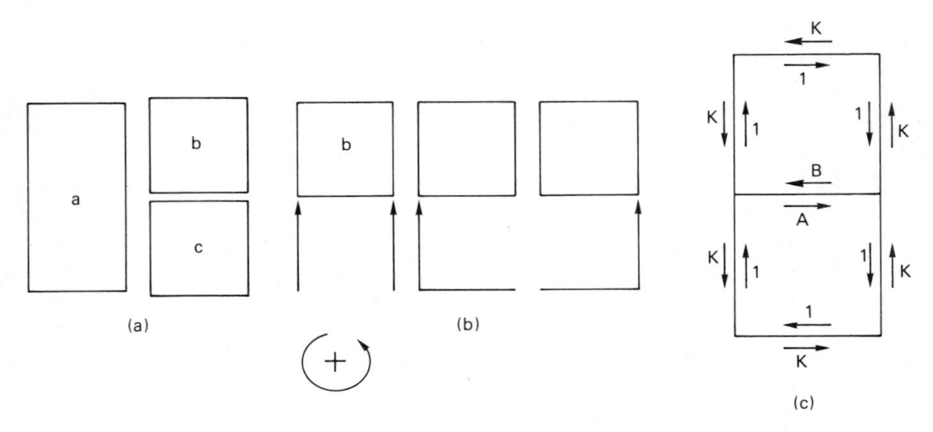

FIG. 2.12 (a) Cycles for Fig. 2.11. (b) Flux diagrams for cycle *b*. (c) Rate constants used in example (1 is the reference rate constant).

cycles, the flux diagrams for cycle *b* (cycle *c* is similar), and the semiassigned rate constants (1 is the reference rate constant; K, A, and B are rate constants relative to the reference). There will be coupling in this model because cycle *a* transports both Na^+ and L. If out → in is to be "downhill" for Na^+, we must have (cycle *b*) $\Pi_{b+} > \Pi_{b-}$, or $K^3A > B$, or $K^3 > B/A$. If out → in is to be uphill for L, we need (cycle *c*) $A > K^3B$, or $A/B > K^3$. If cycle *a* is to transport Na^+ and L out → in, we require $K^6 > 1$. For example, we might have $K = 2$ and $A/B = K^4 = 16$.

This model has seven flux diagrams and $6 \times 15 = 90$ directional diagrams. Also, $f = 7 - 5 = 2$ (Na^+, L).

The cycle fluxes are, from Eq. 2.11,

$$J_a = N(\Pi_{a+} - \Pi_{a-})/\Sigma = N(K^6 - 1)/\Sigma \tag{2.21}$$

$$\begin{aligned} J_b &= N(\Pi_{b+} - \Pi_{b-})(\alpha_{53}\alpha_{64} + \alpha_{65}\alpha_{53} + \alpha_{56}\alpha_{64})/\Sigma \\ &= N(K^3A - B)(K^2 + K + 1)/\Sigma \end{aligned} \tag{2.22}$$

$$\begin{aligned} J_c &= N(\Pi_{c+} - \Pi_{c-})(\alpha_{13}\alpha_{24} + \alpha_{21}\alpha_{13} + \alpha_{12}\alpha_{24})/\Sigma \\ &= N(K^3B - A)(K^2 + K + 1)/\Sigma. \end{aligned} \tag{2.23}$$

We know that Σ has 90 terms, but it is otherwise unknown at this stage.

We choose p_3^∞ and p_4^∞ as the "working" probabilities. For transitions $3 \rightleftarrows 4$ we have, from Eqs. 1.15 and 2.7,

$$J_{34}^\infty \Sigma/N = (J_b\Sigma/N) - (J_c\Sigma/N) = A(p_3^\infty\Sigma) - B(p_4^\infty\Sigma). \tag{2.24}$$

Using Eqs. 2.22 and 2.23 in Eq. 2.24, we have one linear equation in the two unknowns $p_3^\infty\Sigma$ and $p_4^\infty\Sigma$. A second such relation follows, say, from $4 \rightleftarrows 2 \rightleftarrows 1 \rightleftarrows 3$:

$$
\begin{aligned}
J_{42}^\infty \Sigma/N = J_{21}^\infty \Sigma/N = J_{13}^\infty \Sigma/N &= (J_a \Sigma/N) + (J_b \Sigma/N) \\
&= K(p_4^\infty \Sigma) - (p_2^\infty \Sigma) = K(p_2^\infty \Sigma) - (p_1^\infty \Sigma) \\
&= K(p_1^\infty \Sigma) - (p_3^\infty \Sigma).
\end{aligned}
\tag{2.25}
$$

Denoting $p_i^\infty \Sigma \equiv x_i$, we can use $Kx_2 - x_1 = Kx_1 - x_3$ to obtain $Kx_1(x_2, x_3) - x_3$. We can then employ $Kx_4 - x_2 = Kx_1 - x_3$ to find $Kx_4 - x_2(x_3, x_4)$. On setting this equal to $(J_a\Sigma/N) + (J_b\Sigma/N)$, we have the desired second linear relation between $p_3^\infty\Sigma$ and $p_4^\infty\Sigma$. In this way we arrive at

$$
p_3^\infty \Sigma = s(K^3 + 1 + Bs), \qquad p_4^\infty \Sigma = s(K^3 + 1 + As) \tag{2.26}
$$

where $s = K^2 + K + 1$. Then from, say, $p_4^\infty\Sigma$ and (Eq. 2.25)

$$
(J_a \Sigma/N) + (J_b \Sigma/N) = K(p_4^\infty \Sigma) - (p_2^\infty \Sigma),
$$

we get $p_2^\infty\Sigma$. States 1, 5, and 6 are handled in similar fashion. Finally, with all the $p_i^\infty\Sigma$ available, Σ may be deduced by means of $\sum_i p_i = 1$. The result is

$$
\begin{aligned}
p_1^\infty &= s[K^3 + 1 + AK^2 + B(K + 1)]/\Sigma \\
p_2^\infty &= s[K^3 + 1 + AK(K + 1) + B]/\Sigma \\
p_3^\infty &= s(K^3 + 1 + Bs)/\Sigma \\
p_4^\infty &= s(K^3 + 1 + As)/\Sigma \\
p_5^\infty &= s[K^3 + 1 + A + BK(K + 1)]/\Sigma \\
p_6^\infty &= s[K^3 + 1 + A(K + 1) + BK^2]/\Sigma
\end{aligned}
\tag{2.27}
$$

where

$$
\Sigma = 3s[2(K^3 + 1) + (A + B)s]. \tag{2.28}
$$

Note that Σ has 90 terms (with many degeneracies), as expected. Thus, in this example, we have deduced the state probabilities from the very simple flux diagrams (Fig. 2.12), without using directional diagrams.

A numerical example based on these equations will be presented in Chapter 4.

REFERENCES

1. T. L. Hill, *J. Theoret. Biol.* **10**, 442 (1966).
2. T. L. Hill, "Thermodynamics for Chemists and Biologists," Chapter 7. Addison-Wesley, Reading, Massachusetts, 1968.
3. T. L. Hill and Y. Chen, *Proc. Nat. Acad. Sci. U.S.* **72**, 1291 (1975).
4. T. L. Hill, *Biochemistry* **14**, 2127 (1975).
5. T. L. Hill, *Progr. Biophys. Mol. Biol.* **28**, 267 (1974).
6. R. Gordon, *J. Appl. Prob.* **7**, 373 (1970).

Chapter 3 | Fluxes and Forces

The primary objective in the first two chapters was to introduce the diagram method per se. The operational and thermodynamic significance of the formal results obtained was not discussed, except incidentally. In this chapter and in Chapter 4 we turn to considerations of this sort. Roughly speaking, the present chapter is concerned with system properties while Chapter 4 deals with state and transition properties.

The subject of muscle contraction presents special problems so it is reserved for separate treatment in Chapter 5. Also, as mentioned in Section 1.1, light absorbing systems are exceptional and are not properly covered by Chapters 3 and 4. Such systems are discussed briefly in Appendix 5.

The title of this chapter might signify "near equilibrium" to some readers. But this is not the case. As usual, we study steady-state systems arbitrarily far from equilibrium.

We first consider three examples, already introduced, from a thermodynamic point of view. These suffice to bring out the necessary general principles. Following this, we discuss near-equilibrium irreversible thermodynamics as applied to systems of the type being investigated in this book.

3.1 Example: Membrane Transport of Two Ligands

We return to the model introduced in Figs. 1.3–1.5 and 2.1–2.3. Figure 3.1a will serve to remind the reader of the nature of the various transitions (introduced in Fig. 1.3a). An arrow entering the diagram indicates binding to E (e.g., 1A indicates binding of ligand L_1 from bath A). The corresponding

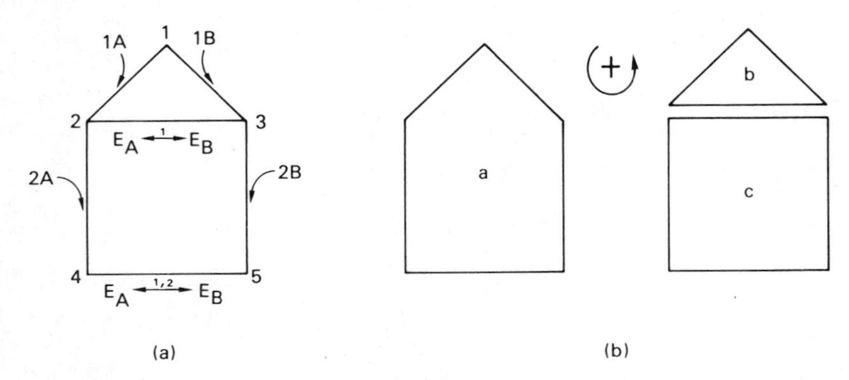

FIG. 3.1 (a) Diagram for transport of ligands 1 and 2 between baths A and B. (b) Cycles.

departing arrows (desorption) are omitted, for simplicity. The horizontal arrows refer to conformational changes in E that switch the bath accessible to bound ligand or ligands (see Fig. 1.3a).

Rate Constants for Binding

Before proceeding with the model, a digression is required to review some properties of rate constants for binding. We begin with the pair α_{12} (binding) and α_{21} (desorption)—see Fig. 3.1a. The desorption rate constant α_{21} will depend primarily on the depth of the binding free energy well from which L_1 must escape in order to return to bath A. This is a true first-order rate constant that is independent of the concentration c_{1A} of L_1 in bath A. On the other hand, the binding rate constant α_{12} will be proportional to c_{1A}—or, if bath A is not dilute in L_1, c_{1A} must be replaced by the concentration activity of L_1. For simplicity, we shall use the symbol c_{1A} to represent either concentration or activity.

The above properties of α_{12}, although well known ("pseudo first-order rate constant"), may be verified by considering the *equilibrium* binding of L_1 from bath A (no other states considered). From detailed balance, we have $\alpha_{12}/\alpha_{21} = p_2^e/p_1^e =$ (number of A-side EL_1)/(number of E). But from thermodynamics or statistical mechanics (1, pp. 125–127), this latter ratio is found to be proportional to the activity of L_1 in bath A (this is essentially the Langmuir adsorption isotherm). Therefore α_{12} is proportional to the activity of L_1 in bath A.

The corresponding rate constants for L_1 and bath B are α_{13} (binding) and α_{31} (desorption), with $\alpha_{13} \sim c_{1B}$. In our examples, we shall assume for simplicity that, when the ensemble is at equilibrium, $c_{1A} = c_{1B}$ and $\alpha_{12} = \alpha_{13}$. This is not the only possibility, but it is the basic case. Exceptions would be, for example: (a) if L_1 is charged and baths A and B are at different electrostatic potentials, then $c_{1A} \neq c_{1B}$ at equilibrium (see Appendix 3); and (b) if there is a free energy barrier to adsorption of L_1 that is different from bath A as compared to bath B, then $\alpha_{12} \neq \alpha_{13}$ even if $c_{1A} = c_{1B}$ at equilibrium (2).

We shall generally assume, then, that $\alpha_{12} = \alpha_1^* c_{1A}$ and $\alpha_{13} = \alpha_1^* c_{1B}$, both with the same proportionality constant α_1^* (α_1^* is, of course, just the second-order rate constant for binding L_1), and that, at equilibrium, $c_{1A} = c_{1B}$. The extent to which L_1 is out of equilibrium in the two baths, in an arbitrary steady state, could then be measured by, for example, $c_{1A} - c_{1B}$ or $\alpha_{12} - \alpha_{13}$ or $\ln \alpha_{12}/\alpha_{13} = \ln c_{1A}/c_{1B}$. This last quantity is closely related to the difference in chemical potential of L_1 in the two baths:

$$\begin{aligned} \mu_{1A} &= \mu_1^0 + kT \ln c_{1A} \\ \mu_{1B} &= \mu_1^0 + kT \ln c_{1B} \end{aligned} \tag{3.1}$$

$$\mu_{1A} - \mu_{1B} = kT \ln(c_{1A}/c_{1B}) \equiv X_1. \tag{3.2}$$

This chemical potential difference is called the "thermodynamic force" X_1 for component 1 (L_1). This force tends to drive L_1 across the membrane in the direction of its decreasing concentration (i.e., $A \rightarrow B$ if $c_{1A} > c_{1B}$ and $X_1 > 0$). At equilibrium, $X_1 = 0$. More conventionally (1), the force is defined as $(\mu_{1A} - \mu_{1B})/T$, but the distinction is not important here since all systems considered are isothermal.

Of course exactly the same comments apply to L_2:

$$\alpha_{24} = \alpha_2^* c_{2A}, \quad \alpha_{35} = \alpha_2^* c_{2B}; \qquad c_{2A}/c_{2B} = \alpha_{24}/\alpha_{35};$$

and the force is

$$\mu_{2A} - \mu_{2B} = kT \ln(c_{2A}/c_{2B}) \equiv X_2. \tag{3.3}$$

As an addendum to this subsection, we make the following comment. We have been considering elementary binding steps above for which a first-order rate constant $\alpha \sim c$. The situation would be more complicated than this if a reduced diagram (Appendix 1) is used in which elementary binding steps are included in the diagram reduction. For example, if k_2 in Eqs. A1.6 refers to an elementary step with $k_2 \sim c$, then both k_+ and k_- would depend on c but neither would be simply proportional to c. Equations A1.13, A1.22, and A1.23 provide other similar examples. This topic has been discussed in more detail in Hill (3), Appendix II, and in the Appendix of Hill *et al.* (4).

Relations Between Rate Constants

Hitherto we have treated the rate constants α_{ij} of a model in a purely formal way without regard to possible interrelations between them. In the models of interest in this book, some (i.e., the pseudo first-order) rate constants—as we have just seen—depend on the concentration of a ligand, substrate, or product, while others would be invariant to concentration changes. Of course *all* rate constants might in general depend on temperature, salt concentration, membrane potential, etc. In any given model, obligatory relations between the rate constants can be found by considering *equilibrium sets* of the concentration dependent rate constants, as illustrated below. The number of such relations will be equal to the number of thermodynamic forces in the model. In the present example, there are two such forces (X_1 and X_2).

At equilibrium, for cycle b (Fig. 3.1b),

$$\Pi_{b+} = \Pi_{b-} \qquad \text{or} \qquad \alpha_{12}^{e}\alpha_{23}\alpha_{31} = \alpha_{21}\alpha_{32}\alpha_{13}^{e}. \tag{3.4}$$

But since $\alpha_{12}^{e} = \alpha_{13}^{e}$ in this model, as discussed above, we have the rate constant connection $\alpha_{23}\alpha_{31} = \alpha_{21}\alpha_{32}$. This is a necessary relation between invariant rate constants of the model that holds under any conditions, not just at equilibrium. It is a consequence, however, of the detailed balance requirement at equilibrium.

At a steady state, then,

$$\begin{aligned}\Pi_{b+} - \Pi_{b-} &= (\alpha_{12} - \alpha_{13})\alpha_{21}\alpha_{32} \\ &= [(\alpha_{12}/\alpha_{13}) - 1]\Pi_{b-} = [(c_{1\text{A}}/c_{1\text{B}}) - 1]\Pi_{b-}.\end{aligned} \tag{3.5}$$

Here, and below, we have arbitrarily chosen bath B as the reference bath. Also, we note that, at a steady state,

$$\begin{aligned}\Pi_{b+}/\Pi_{b-} &= \alpha_{12}/\alpha_{13} = c_{1\text{A}}/c_{1\text{B}} \\ &= e^{(\mu_{1\text{A}} - \mu_{1\text{B}})/kT} = e^{X_1/kT} \\ &= J_{b+}/J_{b-}.\end{aligned} \tag{3.6}$$

The last relation follows from Eq. 2.13. This illustrates the fact that there is always a direct connection (via $\Pi_{\kappa+}/\Pi_{\kappa-}$) between the thermodynamic force acting in a cycle κ of a diagram and the ratio of one-way cycle fluxes $J_{\kappa+}/J_{\kappa-}$.

The corresponding relations for cycle c are

$$\begin{aligned}\alpha_{45}\alpha_{53}\alpha_{32} &= \alpha_{42}\alpha_{54}\alpha_{23} \\ \Pi_{c+} - \Pi_{c-} &= (\alpha_{24} - \alpha_{35})\alpha_{42}\alpha_{54}\alpha_{23} \\ &= [(c_{2\text{A}}/c_{2\text{B}}) - 1]\Pi_{c-}\end{aligned} \tag{3.7}$$

$$\begin{aligned}\Pi_{c+}/\Pi_{c-} &= \alpha_{24}/\alpha_{35} = c_{2\text{A}}/c_{2\text{B}} \\ &= e^{(\mu_{2\text{A}} - \mu_{2\text{B}})/kT} = e^{X_2/kT} = J_{c+}/J_{c-}.\end{aligned} \tag{3.8}$$

Cycle a adds nothing new in the way of required rate constant relationships since there are only two independent forces. Thus, from $\Pi_{a+} = \Pi_{a-}$ at equilibrium, we find $\alpha_{45}\alpha_{53}\alpha_{31} = \alpha_{21}\alpha_{42}\alpha_{54}$, which is just a combination of the corresponding expressions from cycles b and c. Also, we have

$$\begin{aligned} \Pi_{a+} - \Pi_{a-} &= (\alpha_{12}\alpha_{24} - \alpha_{13}\alpha_{35})\alpha_{21}\alpha_{42}\alpha_{54} \\ &= [(c_{1A}c_{2A}/c_{1B}c_{2B}) - 1]\Pi_{a-} \end{aligned} \tag{3.9}$$

$$\begin{aligned} \Pi_{a+}/\Pi_{a-} &= c_{1A}c_{2A}/c_{1B}c_{2B} = e^{(X_1+X_2)/kT} \\ &= J_{a+}/J_{a-} . \end{aligned} \tag{3.10}$$

Operational Fluxes

The thermodynamic forces introduced above are operational or observable. They are determined by ligand bath concentrations. We turn now to the corresponding thermodynamic or operational fluxes (1, 5). Our object here is to give explicit expressions (except for Σ, which contains 55 terms) for these fluxes at an arbitrary steady state. The independent variables are c_{1A} and c_{2A}, or X_1 and X_2 (c_{1B} and c_{2B}—in the reference bath—are considered constant).

There is a steady-state flux, between the baths and across the membrane, in the two components (ligands) 1 and 2. We denote these fluxes by J_1^∞ and J_2^∞ in the direction A → B. The rate at which molecules of component 1 leave bath A is J_{12}^∞. Therefore $J_1^\infty = J_{12}^\infty$. Similarly, $J_2^\infty = J_{24}^\infty$. Thus the steady-state fluxes J_1^∞ and J_2^∞ are given by (Eq. 2.5 and 2.8)

$$\begin{aligned} J_1^\infty \Sigma/N &= \text{cycle } a + \text{cycle } b, \\ J_2^\infty \Sigma/N &= \text{cycle } a + \text{cycle } c. \end{aligned} \tag{3.11}$$

We digress to note that the maximum possible number of operational fluxes is given by f (Section 1.5), the number of independent transition fluxes. In this case $f = 6 - 4 = 2$.

The forces that produce the steady-state fluxes J_1^∞ and J_2^∞ in this model are X_1 and X_2. As we have just seen, the first of these occurs in cycle b and the second in cycle c. *Both* forces are operative in cycle a; it is this feature that tells us, without benefit of explicit algebra, that there will be thermodynamic coupling and hence free energy transduction (details below) between the two forces and fluxes. Since a cycle makes the same contribution to the flux (in the same direction) along all lines making up the cycle (Section 2.1), either force in cycle a will cause a flux in *both* components. Furthermore, the words "same contribution" used above lead to the near-equilibrium "reciprocal relation," as we shall see in Section 3.4.

In general, any model whose diagram contains one or more cycles that involve two or more forces will show coupling between the forces and fluxes. A more complicated example, illustrating Na^+, K^+ active transport (Fig. 2.7), will be discussed in the next section.

It is conventional—and we have followed this convention above—to define a thermodynamic force X_i and the corresponding thermodynamic flux J_i^∞ in such a way that J_i^∞ has the same sign as X_i *when* X_i *is the only force operative*. Thus, in the current model, if $c_{2B} > c_{2A}$ (X_2 is negative) and $c_{1A} = c_{1B}$ ($X_1 = 0$), J_2^∞ will be negative (L_2 will move in the direction $B \rightarrow A$). But, even with this convention, X_2 and J_2^∞ need not have the same sign if $X_1 \neq 0$ (see below).

Returning now to Eq. 3.11, using Eqs. 2.10, 2.12, and 3.5–3.10, we have

$$\text{cycle } a = [e^{(X_1+X_2)/kT} - 1]a, \qquad a = \Pi_{a-} \tag{3.12}$$

$$\text{cycle } b = (e^{X_1/kT} - 1)b, \qquad b = \Pi_{b-}\Sigma_b \tag{3.13}$$

$$\text{cycle } c = (e^{X_2/kT} - 1)c, \qquad c = \Pi_{c-}\Sigma_c\,. \tag{3.14}$$

Then, to separate the forces, we introduce the identity

$$e^{(X_1+X_2)/kT} - 1 = (e^{X_1/kT} - 1)(e^{X_2/kT} - 1) + (e^{X_1/kT} - 1) + (e^{X_2/kT} - 1)$$

in Eq. 3.12. Also, because Σ_c contains α_{12}, which is proportional to the variable c_{1A}, we rewrite c in Eq. 3.14 as

$$c = [(e^{X_1/kT} - 1) + 2]c', \qquad c' = \alpha_{13}\Pi_{c-}\,. \tag{3.15}$$

Equations 3.11 for the fluxes then become

$$J_1^\infty\Sigma/N = (e^{X_1/kT} - 1)(a+b) + (e^{X_2/kT} - 1)a + (e^{X_1/kT} - 1)(e^{X_2/kT} - 1)a \tag{3.16}$$

$$J_2^\infty\Sigma/N = (e^{X_1/kT} - 1)a + (e^{X_2/kT} - 1)(a+2c') + (e^{X_1/kT} - 1)(e^{X_2/kT} - 1)(a+c'). \tag{3.17}$$

These are the fluxes at an arbitrary steady state, not necessarily near equilibrium. The coefficients (all positive) of the force expressions on the right-hand sides are properties of the steady state, not of an equilibrium state. However, these coefficients are constants if the forces are varied only by changing c_{1A} and c_{2A} (i.e., if c_{1B} and c_{2B}—in the reference bath—are held constant). Note further that α_{12} and α_{24} are included in Σ (i.e., reference bath B has not been introduced in Σ as it has been on the right-hand sides of the above equations). Note also that even in the above form the coefficients of linear "force" terms obey a reciprocal relation. That is, the coefficient $a = \Pi_{a-}$ of $e^{X_2/kT} - 1$ in Eq. 3.16 is equal to the coefficient of $e^{X_1/kT} - 1$ in Eq. 3.17. See Section 3.4 in this connection.

Of course, near equilibrium (X_1 and X_2 small), we can write

$$\begin{aligned} e^{X_1/kT} - 1 &= (X_1/kT) + \tfrac{1}{2}(X_1/kT)^2 + \cdots \\ e^{X_2/kT} - 1 &= (X_2/kT) + \tfrac{1}{2}(X_2/kT)^2 + \cdots \end{aligned} \tag{3.18}$$

(see Section 3.4). It might be recalled here that, in terms of concentrations,

$$e^{X_1/kT} - 1 = (c_{1A}/c_{1B}) - 1, \qquad X_1/kT = \ln(c_{1A}/c_{1B}), \qquad \text{etc.}$$

Thermodynamic Remarks

We verify first that, because of the coupling in this model, active transport is possible. Let us examine the conditions under which the $A \to B$ gradient in L_1, i.e., $c_{1A} > c_{1B}$ and $X_1 > 0$, can drive L_2 uphill ($A \to B$) *against* its own concentration gradient ($c_{2B} > c_{2A}$, $X_2 < 0$). In other words, we are interested in conditions such that J_2^∞ is positive ($A \to B$) even though X_2 is negative. First we note, from Eq. 3.17, that if $X_2 = 0$ (i.e., $c_{2A} = c_{2B}$),

$$J_2^\infty \Sigma/N = (e^{X_1/kT} - 1)a > 0. \tag{3.19}$$

Thus $J_2^\infty > 0$ if $X_2 = 0$: component 2 is transported $A \to B$ by the component 1 gradient. If X_2 is made negative (by decreasing c_{2A} below the fixed value of c_{2B}), J_2^∞ will remain positive in sign but decrease in magnitude until $J_2^\infty = 0$ is reached when X_2 has the value X_2' determined by (Eq. 3.17)

$$e^{X_2'/kT} - 1 = \frac{-a(e^{X_1/kT} - 1)}{a + 2c' + (a + c')(e^{X_1/kT} - 1)}. \tag{3.20}$$

Thus transport of L_2 *uphill* by the gradient occurs, for a given $X_1 > 0$, when $0 > X_2 > X_2'$. If the only important cycle is a, i.e., if $c' = 0$, Eq. 3.20 gives $X_2' = -X_1$ as expected.

In this example, when $0 < X_2 < X_2'$ is satisfied, the rate of free energy decrease in L_1 is $J_1^\infty X_1$ while the rate of gain in free energy by L_2 is $J_2^\infty(-X_2)$. The former quantity is necessarily larger than the latter (see below). The efficiency η of free energy conversion or transduction from L_1 to L_2 in this case is thus

$$\eta = J_2^\infty(-X_2)/J_1^\infty X_1. \tag{3.21}$$

The net steady-state rate of free energy dissipation or of entropy production (1) in the ensemble + bath A + bath B is

$$\begin{aligned} T\, d_iS/dt &= J_1^\infty X_1 + J_2^\infty X_2 \\ &= J_1^\infty X_1 - J_2^\infty(-X_2) > 0. \end{aligned} \tag{3.22}$$

(Strictly, for the rate of entropy production, we have to multiply by $1/T$.) The *net* positive sign here is required by the second law of thermodynamics

(1). Therefore $J_1^\infty X_1 > J_2^\infty(-X_2)$ and $0 \leq \eta \leq 1$ as expected. Only *part* of the free energy $J_1^\infty X_1$ is transformed into $J_2^\infty(-X_2)$; the rest is dissipated as heat [ensemble + bath A + bath B are in contact with a constant-temperature heat bath (1)].

The relations 3.19–3.22 apply to steady states arbitrarily far from equilibrium. With a given set of rate constants and forces (bath concentrations) for the model, all of these quantities could of course be calculated explicitly.

General Comments

Although this is an arbitrary and simple model of free energy transduction in biology, it can serve as a prototype for the following quite general comments.

(1) The model (Fig. 1.3a) makes use only of adsorption–desorption of ligands and of protein conformational changes (other models include electrostatic effects, chemical reactions, light absorption, electron transfer, and—in muscle contraction—relative translational motion). Only conventional physical processes are involved and only conventional statistical physics is required for the analysis of these processes.

(2) The free energy transduction accomplished is a property of the *entire model and diagram*—*all* transitions are involved. The free energy transfer process *cannot* be ascribed to a *single* critical step or transition as is sometimes suggested in qualitative arguments, though a single step might have some kind of unique role (e.g., light absorption).

(3) Each particular problem in bioenergetic free energy transduction will of course introduce its own physical processes and special complications—many of which are undoubtedly not known yet. This temporary ignorance naturally leads to rival speculations and hence to controversy. But as far as *general principles* are concerned, free energy transduction in biology does not appear to be a mysterious phenomenon.

(4) Although the bioenergetics of muscle contraction has unusual and complicating features (Chapter 5), it is possibly the best understood of these problems at the present time. This is because of the wide variety of experimental methods that can be and have been used. But the quite recently discovered phototranslocation system in *H. halobium* (6, 7) is very much simpler and will probably be the first to be worked out in complete detail.

(5) Is there a need (as suggested by David E. Green) for a new unifying principle of bioenergetics? The present writer's point of view is that a "unifying principle" has in fact been with us a long time: it is a belief in the applicability of the ordinary laws of statistical physics, including the second law of thermodynamics, introduced as required *in each special case*. In the

context of the present book, "introduced as required" means construction of the molecular model and of the corresponding diagram—with a full set of self-consistent rate constants.

Cycles and Entropy Production

Equation 3.22 gives the rate of entropy production as a sum of products of operational forces and fluxes, $J_1^\infty X_1 + J_2^\infty X_2$. The sum must be positive but the separate terms in the sum need not always be positive. The entropy production can be expressed in several other ways, and we consider one of these alternatives here (see also Chapters 4 and 5).

Instead of collecting terms according to the thermodynamic forces acting in the model or diagram, as in Eq. 3.22, let us collect them according to the *cycles* of the diagram. Since (Eq. 2.8)

$$J_1^\infty = J_{12}^\infty = J_a + J_b \qquad \text{and} \qquad J_2^\infty = J_{24}^\infty = J_a + J_c,$$

we have

$$\begin{aligned} T\, d_i S/dt &= (J_a + J_b)X_1 + (J_a + J_c)X_2 \\ &= J_a X_a + J_b X_b + J_c X_c > 0 \end{aligned} \tag{3.23}$$

where

$$X_a \equiv X_1 + X_2, \qquad X_b \equiv X_1, \qquad X_c \equiv X_2. \tag{3.24}$$

This same rearrangement can be carried out for any model, and we shall encounter further examples below and in Chapter 4. The coefficient X_κ of each cycle flux J_κ will obviously be the sum of all thermodynamic forces operative in cycle κ. Some cycles in some models may contain no forces, or no net force. For an arbitrary model,

$$T\, d_i S/dt = \sum_\kappa J_\kappa X_\kappa = J_a X_a + J_b X_b + \cdots > 0. \tag{3.25}$$

Furthermore, as we have seen in Eqs. 2.11, 2.13, and 3.4–3.10 (and as will be more obvious in terms of the free energy levels of states to be introduced in the next chapter—see Eqs. 4.16–4.18), both J_κ amd X_κ are directly related to the rate constant products $\Pi_{\kappa+}$ and $\Pi_{\kappa-}$ for cycle κ:

$$\begin{aligned} e^{X_\kappa/kT} &= \Pi_{\kappa+}/\Pi_{\kappa-} \\ J_\kappa &= N(\Pi_{\kappa+} - \Pi_{\kappa-})\Sigma_\kappa/\Sigma. \end{aligned} \tag{3.26}$$

Since $\Pi_{\kappa\pm}$, Σ_κ, and Σ are positive quantities, J_κ and X_κ will always have the same sign. Also, if a cycle κ has no net force, i.e., if $X_\kappa = 0$, then it follows that the steady-state $J_\kappa = 0$ as well. Thus, in Eq. 3.25, not only is the sum positive but *each term* in the sum is positive (or zero): each cycle makes an additive positive (or zero) contribution to the total rate of entropy production in the ensemble. The same cannot be said about Eq. 3.22. The difference arises on passing to the more detailed and mechanistic (nonoperational) subdivision of terms used in Eqs. 3.23 and 3.25. A still finer subdivision of terms, i.e., on the basis of *individual* transition pairs (8, 9) (the *lines* of a diagram), will be introduced in Chapter 4, and this same property (nonnegative terms) will be seen to persist. Keizer (10) has discussed this latter case in a more general context.

3.2 Substrate–Product Rate Constant Relations

Our next main objective in this chapter is to discuss the more complicated model introduced in Fig. 2.7. The chemical reaction ATP → ADP + P_i, catalyzed by the Na,K-ATPase of the hypothetical model, is involved. As a prerequisite, therefore, we first digress in this section to examine rate constant relations for an arbitrary reaction S ⇄ P catalyzed by an enzyme. This supplements our adsorption–desorption discussion above and will suffice to provide the necessary background.

We consider explicitly the case in Fig. 1.1. The rate constant α_{12} relates to the binding of S to E while α_{13} refers to the binding of P to E. As in Section 3.1, both can be written in the form $\alpha_{12} = \alpha_{12}^* c_S$ and $\alpha_{13} = \alpha_{13}^* c_P$, where c_S and c_P are solution concentrations and α_{12}^* and α_{13}^* are second-order binding rate constants. We regard c_P as a constant while c_S might have different values in different experiments. (This choice is obviously arbitrary and could as well be reversed, or both concentrations could vary.) In particular, c_S^e represents the concentration of substrate that would be in equilibrium with P at concentration c_P.

The chemical potentials of S and P in solution are, in standard notation,

$$\mu_S = \mu_S^0 + kT \ln c_S , \qquad \mu_P = \mu_P^0 + kT \ln c_P . \tag{3.27}$$

At equilibrium,

$$\mu_S^e = \mu_S^0 + kT \ln c_S^e = \mu_P = \mu_P^0 + kT \ln c_P \tag{3.28}$$

so that

$$c_P / c_S^e = e^{-(\mu_P{}^0 - \mu_S{}^0)/kT} \equiv K. \tag{3.29}$$

This is the conventional relation between the equilibrium concentration (mass action) quotient and the standard free energy change, $\mu_P^0 - \mu_S^0$, for the reaction S → P. K is the corresponding equilibrium constant.

There is a positive drive or thermodynamic force X for the turnover of S into P by the enzyme E when $c_S > c_S^e$; that is, when $\mu_S > \mu_S^e$ or $\mu_S > \mu_P$ (recall that $\mu_S^e = \mu_P$). The magnitude of the force is $\mu_S - \mu_P$ or $\mu_S - \mu_S^e$, which is related to concentrations by

$$X = \mu_S - \mu_P = kT \ln(c_S/c_S^e). \quad (3.30)$$

This is the analogue of Eqs. 3.2 and 3.3. As it refers to a chemical reaction, X is often called an "affinity."

Turning now to the rate constants in the cycle (Fig. 1.1), at equilibrium $\Pi_+ = \Pi_-$ and we have

$$\alpha_{12}^e \alpha_{23} \alpha_{31} = \alpha_{21} \alpha_{32} \alpha_{13} \quad (3.31)$$

or

$$\alpha_{12}^* c_S^e \alpha_{23} \alpha_{31} = \alpha_{21} \alpha_{32} \alpha_{13}^* c_P . \quad (3.32)$$

Equation 3.31 or 3.32 is the required relationship that the rate constants of the cycle must satisfy. At an arbitrary steady state (i.e., arbitrary c_S),

$$\frac{\Pi_+}{\Pi_-} = \frac{\alpha_{12}\alpha_{23}\alpha_{31}}{\alpha_{21}\alpha_{32}\alpha_{13}} = \frac{\alpha_{12}}{\alpha_{12}^e} = \frac{c_S}{c_S^e} = e^{X/kT}. \quad (3.33)$$

This is closely analogous to Eqs. 3.6 and 3.8. Thus this chemical reaction can be treated formally in the same way as a concentration gradient. We note also, from Eqs. 3.29 and 3.32, that

$$K = \frac{c_P}{c_S^e} = e^{-(\mu_P{}^0 - \mu_S{}^0)/kT} = \frac{\alpha_{12}^* \alpha_{23} \alpha_{31}}{\alpha_{21} \alpha_{32} \alpha_{13}^*}. \quad (3.34)$$

This rate constant quotient is invariant (unlike Eq. 3.33) in the sense that it does not depend on the concentrations of S and P. This last result will be needed in the next chapter but Eq. 3.33 is the important relationship for present purposes.

The ratio c_S/c_S^e might be very large in some cases (i.e., c_S^e might be extremely small—essentially zero). For example, if $\mu_S - \mu_P = 12.5$ kcal mol^{-1} at 20°C (the order of magnitude for ATP → ADP + P_i), we would have, according to Eq. 3.30, $c_S/c_S^e = 2.07 \times 10^9$. Note that the free energy difference here is *not* the standard free energy decrease $\mu_S^0 - \mu_P^0$ but the *actual* free energy decrease.

The flux for this single-cycle model (Fig. 1.1) is (Eq. 1.20)

$$J^\infty = N[(c_S/c_S^e) - 1]\Pi_-/\Sigma. \quad (3.35)$$

Σ has nine terms (Eqs. 1.10–1.12), three of which contain α_{12}. Hence Σ is linear in c_S. Thus J^∞ has a concentration dependence of the form $(Ac_S - B)/(Cc_S + D)$ (recall that we are arbitrarily keeping c_P constant here).

For large c_S, J^{∞} reaches the limiting (saturation) value A/C. Equation 3.35 is, of course, nothing new. It is a well-known generalization (2, 11), in different notation, of the Michaelis–Menten equation.

Charged Ligand

For a ligand (uncharged) distributed across the membrane (Section 3.1), equilibrium occurs when $c_A = c_B$. For the reaction S ⇄ P, equilibrium does *not* correspond to $c_S = c_P$ so we have introduced the alternative procedure of comparing c_S with c_S^e (since $c_S = c_S^e$ at equilibrium). This latter procedure can be adopted as well for ligand transport when $c_A \neq c_B$ at equilibrium. Thus, for a ligand with charge $z\varepsilon$ (Appendix 3), if bath B (the reference bath) is at an electrostatic potential $\psi = 0$ while bath A is at potential ψ, at equilibrium we would have $c_A^e e^{z\varepsilon\psi/kT} = c_B$ since the ligand electrochemical potential must be the same in the two phases. At an arbitrary steady state (c_B constant, $c_A \neq c_A^e$), the ratio c_A/c_A^e has the same significance and utility as c_S/c_S^e above.

Release of Product in Steps

Equations 3.31 and 3.33 are essentially unaffected when the product (P) is released in two or more steps. Figure 3.2 provides an example. Although there are now three binding steps, as indicated in Fig. 3.2, if c_{ADP} and c_{P_i} are held constant and only c_{ATP} is allowed to vary, Eqs. 3.31 and 3.33 are replaced by

$$\alpha_{12}^e \alpha_{23} \alpha_{34} \alpha_{41} = \alpha_{21} \alpha_{32} \alpha_{43} \alpha_{14} \tag{3.36}$$

and

$$\frac{\Pi_+}{\Pi_-} = \frac{\alpha_{12}\alpha_{23}\alpha_{34}\alpha_{41}}{\alpha_{21}\alpha_{32}\alpha_{43}\alpha_{14}} = \frac{\alpha_{12}}{\alpha_{12}^e} = \frac{c_{ATP}}{c_{ATP}^e} = e^{X/kT}, \tag{3.37}$$

where

$$X = \mu_{ATP} - (\mu_{ADP} + \mu_{P_i}). \tag{3.38}$$

Because of these results, for simplicity we shall generally treat the release of ATP products as if this occurred in a single step (e.g., as if state 4 in Fig. 3.2 were a transient intermediate).

Successive Reactions with Common Intermediate

We have introduced (by means of an example), in Section 3.1, the concepts of thermodynamic coupling and free energy transduction within a diagram. In the present section, we have seen how to include chemical reactions in the formalism.

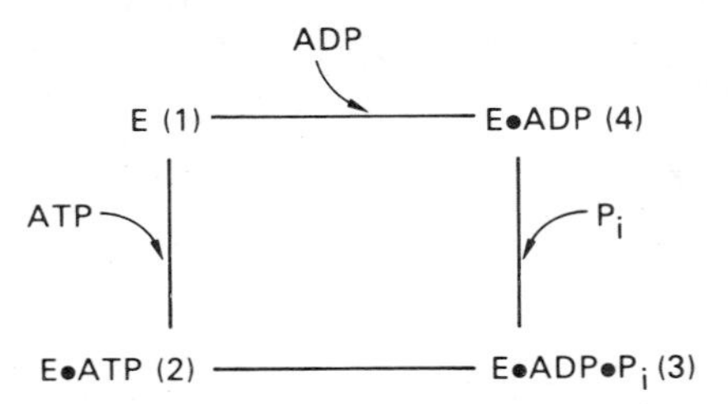

FIG. 3.2 Cycle illustrating release of products (P_i, ADP) in steps (counterclockwise direction).

We digress here to remark that the most common type of free energy transduction in biochemistry involves two or more successive chemical reactions, with common intermediates, each reaction catalyzed by its own enzyme. Two examples are

(a) $S \to P_1 \quad (E_1), \qquad P_1 \to P_2 \quad (E_2);$
(b) $A + B \to C + D \quad (E_1),$
$\quad C + F \to G + H \quad (E_2) \qquad$ (e.g., C = ATP).

The common intermediates are P_1 and C, respectively. From the point of view of this book, each separate reaction here is an independent problem and has its own independent diagram. The coupling and transduction we deal with explicitly, in our examples, is *within* a *single* diagram.

3.3 Example: Active Transport of Na^+ and K^+

We return here to the hypothetical model introduced briefly in Section 2.3 and Figs. 2.6–2.9. The earlier discussion should be reviewed. In a formal way this model is very similar to one discussed elsewhere (1, 5) at some length, but the meaning of the states has been altered here to conform more closely to current views (12).

We arbitrarily choose B (outside) as the reference bath, and select Na^+ as component 1 and K^+ as component 2. The thermodynamic fluxes and forces J_1^∞, J_2^∞, X_1, and X_2 are all taken as positive (Fig. 3.3) in the direction A(in) → B(out). Note, in Fig. 3.3, that $J_2^\infty > 0$ corresponds to clockwise circulation. For ATP, the overall reaction is ATP(in) → ADP(in) + P_i(out), since the enzyme E is phosphorylated following binding of ATP from the inside and then P_i is released later on the outside. J_T^∞ and X_T are taken as positive in the direction ATP → products.

This model, also (Section 3.1), employs the maximum number of operational fluxes: $f = 10 - 7 = 3$.

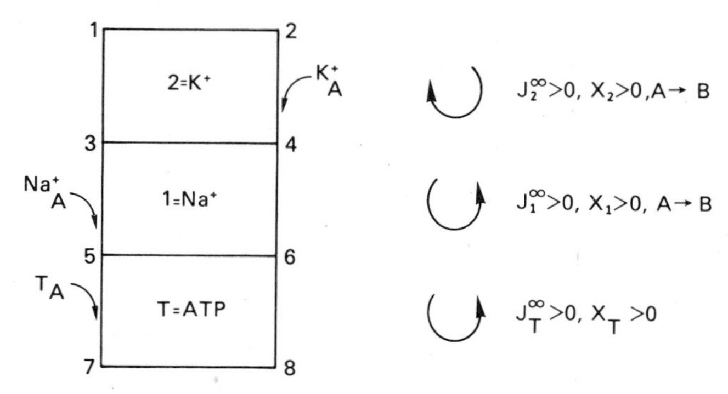

FIG. 3.3 Diagram for hypothetical Na,K-ATPase model (Fig. 2.6). Flux-force conventions are shown.

The forces are defined by (see Sections 3.1 and 3.2)

$$X_1 = \mu_{1A} - \mu_{1B} = kT \ln(c_{1A}/c_{1B}) \tag{3.39}$$

$$X_2 = \mu_{2A} - \mu_{2B} = kT \ln(c_{2A}/c_{2B}) \tag{3.40}$$

$$X_T = \mu_T - \mu_D = \mu_T - \mu_T^e = kT \ln(c_T/c_T^e), \tag{3.41}$$

where D = ADP, P_i and $\mu_D = \mu_{ADP} + \mu_{P_i}$. Because this model is primarily pedagogical and already quite complicated, we ignore the fact that Na^+ and K^+ are charged and that there is a membrane potential. These features would, in fact, make only trivial differences at the formal level to be used below. Appendix 3 treats an explicit, simpler example that emphasizes these complications.

The usual situation, in active transport of Na^+ and K^+, is as follows: for Na^+, $c_{1B}(\text{out}) > c_{1A}(\text{in})$ and hence $X_1 < 0$; for K^+, $c_{2A}(\text{in}) > c_{2B}(\text{out})$ and hence $X_2 > 0$; and for ATP, $X_T > 0$. If there were no coupling between these three processes, the fluxes would all have the same signs as the corresponding forces: $J_1^\infty < 0$ (Na^+ out → in); $J_2^\infty > 0$ (K^+ in → out); and $J_T^\infty > 0$ (T → D). Actually, X_T is large enough and the coupling strong enough to achieve a *reversal of the signs* of J_1^∞ and J_2^∞: $J_1^\infty > 0$ (Na^+ in → out) and $J_2^\infty < 0$ (K^+ out → in). This is the "active transport." That is, the free energy of T → D is used to transport both Na^+ and K^+ across the cell membrane *against* their electrochemical gradients (expressed here as a simple concentration gradient).

General Comments

The present model furnishes a hypothetical and illustrative scheme to accomplish this transfer of chemical free energy (X_T) into transport free energy (X_1, X_2). Even though the detailed model is not to be taken seriously, it is a reasonable prototype and will suffice to illustrate all the *general* principles involved. In particular, and this point is worth emphasizing in advance (see also Section 3.1), there is no *single* critical or crucial step at which the chemical free energy is suddenly "converted" into transport free energy. Rather, the conversion is accomplished by (ordinary) steady-state kinetic use of *all* the (ordinary) states in the diagram, with the direction of net cycling imposed by the dominating force of $T \to D$, $X_T > 0$. Note, though, that the chemical reaction must have a vectorial aspect in order to accomplish a vectorial result (transport). The remaining problem—and it is of course much more interesting and difficult than the formal generalities discussed here—is to find the actual molecular structures, states, and mechanism involved in the workings of the Na,K-ATPase, i.e., to find the *real* molecular model.

The involvement of *all* the states in the free energy transfer process will be reinforced in the next chapter where we discuss the relative free energy levels of the states of a diagram.

Free Energy Conversion

The net rate at which Na^+ free energy increases is $J_1^\infty(-X_1)$. The same quantity for K^+ is $(-J_2^\infty)X_2$. The rate at which ATP free energy is lost is $J_T^\infty X_T$. All six of the factors (including the negative signs) in these three products are positive quantities in ordinary active transport. (The process could be reversed to synthesize ATP.) The net steady-state rate of dissipation of free energy, or the rate of entropy production (in ensemble + bath A + bath B), is

$$\begin{aligned} T\, d_iS/dt &= J_1^\infty X_1 + J_2^\infty X_2 + J_T^\infty X_T > 0 \\ &= J_T^\infty X_T - [J_1^\infty(-X_1) + (-J_2^\infty)X_2] > 0. \end{aligned} \tag{3.42}$$

The efficiency of steady-state free energy conversion from ATP to Na^+ and K^+ transport is then obviously

$$\eta = \frac{J_1^\infty(-X_1) + (-J_2^\infty)X_2}{J_T^\infty X_T} < 1, \tag{3.43}$$

where both numerator and denominator are positive quantities.

Let us digress to estimate the experimental value of η. It is currently believed that, for each molecule of ATP split, three Na^+ and two K^+ are transported across the membrane. In this case we would have

$$\eta = [3(-X_1) + 2X_2]/X_T\,. \tag{3.44}$$

If the Na^+ equilibrium potential is estimated (for squid giant axon membrane, say) to be 115 mV more positive than the membrane rest potential and the K^+ equilibrium potential is taken as 10 mV more negative than the rest potential, the numerator in Eq. 3.44 becomes $3 \times 2.65 + 2 \times 0.23 \cong 8.42$ kcal mol^{-1}. If $X_T \cong 12.5$ kcal mol^{-1}, the efficiency η is about 67%.

Cycle Fluxes and Forces

We return now to our particular model, Figs. 2.6–2.9 and 3.3. The six cycle fluxes J_κ (Fig. 2.7) are all taken as positive in the counterclockwise direction. Then the cycle contributions to the thermodynamic fluxes are

$$\begin{aligned} J_1^\infty &= J_{35}^\infty = J_{64}^\infty = J_b + J_d + J_f + J_h \\ J_2^\infty &= J_{24}^\infty = J_{12}^\infty = J_{31}^\infty = -J_a - J_d - J_h \\ J_T^\infty &= J_{57}^\infty = J_{78}^\infty = J_{86}^\infty = J_c + J_f + J_h. \end{aligned} \tag{3.45}$$

Note that if cycle h is the only significant cycle, $J_1^\infty = -J_2^\infty = J_T^\infty = J_h$. That is, there is *complete coupling* between the three fluxes. In this case the efficiency is $\eta = (-X_1 + X_2)/X_T$. Of course, there is *always* complete coupling when a single cycle model has more than one flux (and force).

As already emphasized, the important case is $X_1 < 0$, $X_2 > 0$, and $X_T > 0$. Furthermore, X_T is large enough to "overpower" the other two forces combined: $X_T + X_1 - X_2 > 0$. Let us look at the signs of each of the J_κ and X_κ in this case. If we substitute Eqs. 3.45 into Eq. 3.42, we find for the entropy production, subdivided by cycles (see Eq. 3.24),

$$T\, d_iS/dt = \sum_\kappa J_\kappa X_\kappa, \tag{3.46}$$

where

$$\begin{aligned} &X_a = -X_2, \quad X_b = X_1, \quad X_c = X_T, \quad X_d = X_1 - X_2, \\ &X_f = X_1 + X_T, \quad X_h = X_1 - X_2 + X_T. \end{aligned} \tag{3.47}$$

Equations 3.47 are all obvious, as well, directly from Figs. 2.7 and 3.3. It will be recalled that J_κ and X_κ always have the same sign. From Eqs. 3.47 and the assumed signs of X_1, X_2, X_T just mentioned, we see that the signs of *both* J_κ and X_κ in this example are

$$a = -, \quad b = -, \quad c = +, \quad d = -, \quad f = +, \quad h = +.$$

The dominant direction is + (counterclockwise); the dominant cycles are c, f, and h (all involve X_T); cycles a, b, and d contribute clockwise (−) circulation but are of secondary importance. The respective signs of the separate terms and of their sums in Eqs. 3.45 for the thermodynamic fluxes are

$$\begin{aligned} J_1^\infty &= -,\ -,\ +,\ + > 0 \quad \circlearrowleft \\ J_2^\infty &= +,\ +,\ - < 0 \quad \circlearrowleft \\ J_T^\infty &= +,\ +,\ + > 0 \quad \circlearrowleft \end{aligned}$$

All net thermodynamic fluxes are counterclockwise, as indicated above (because of the dominance of X_T); compare Fig. 3.3.

Thermodynamic Fluxes and Forces

The separate cycle fluxes can be written

$$J_a = -N(e^{X_2/kT} - 1)\Pi_{a+}\Sigma_a/\Sigma \tag{3.48}$$

$$J_b = N(e^{X_1/kT} - 1)\Pi_{b-}\Sigma_b/\Sigma \tag{3.49}$$

$$J_c = N(e^{X_T/kT} - 1)\Pi_{c-}\Sigma_c/\Sigma \tag{3.50}$$

$$J_d = N(\Pi_{d+} - \Pi_{d-})\Sigma_d/\Sigma \tag{3.51}$$

$$J_f = N[e^{(X_1+X_T)/kT} - 1]\Pi_{f-}\Sigma_f/\Sigma \tag{3.52}$$

$$J_h = N(\Pi_{h+} - \Pi_{h-})/\Sigma, \tag{3.53}$$

where the number of terms in each of $\Sigma_a, \Sigma_b, \ldots, \Sigma_f$ is indicated in Fig. 2.7 (see also Fig. 2.8), and the number of terms in Σ is $8 \times 56 = 448$ (Section 2.3).

The situation here is more complicated than in Section 3.1 and Fig. 3.1. There the variable (bath A) concentration dependent rate constants are α_{12} and α_{24}. These do not occur at all in the $\Pi_{\kappa-}$ (Eqs. 3.12–3.14) but Σ_c contains α_{12} while Σ contains both α_{12} and α_{24}. We replaced α_{12} by X_1 in Σ_c, by means of Eq. 3.15, but we left Σ unaltered (to avoid tedious algebra).

Here we have to retreat a step, so to speak (again to avoid undue algebraic complexity). The variable (bath A) concentration dependent rate constants are α_{24}, α_{35}, and α_{57} (Fig. 3.3). One or more of these occur in each of $\Sigma_a, \Sigma_b, \Sigma_c$, and Σ_f. Therefore we leave Σ *and all the* Σ_κ unaltered and express *only* the $\Pi_{\kappa+} - \Pi_{\kappa-}$ in terms of the thermodynamic forces X_i. This has already been done in Eqs. 3.48–3.53 for cycles *a, b, c,* and *f* since Π_{a+}, Π_{b-}, Π_{c-}, and Π_{f-} do not contain α_{24}, α_{35}, or α_{57}. But cycles *d* and *h* require separate treatment since both Π_{d-} and Π_{h-} contain α_{24} as a factor. We write this factor as

$$\alpha_{24} = \alpha_{13}\{[(\alpha_{24}/\alpha_{13}) - 1] + 1\},$$

where $\alpha_{24}/\alpha_{13} = e^{X_2/kT}$. Then Eqs. 3.51 and 3.53 become

$$J_d = N[(e^{X_1/kT} - 1) - (e^{X_2/kT} - 1)]\Pi'_d\Sigma_d/\Sigma \tag{3.54}$$

$$J_h = N\{[e^{(X_1+X_T)/kT} - 1] - (e^{X_2/kT} - 1)\}\Pi'_h/\Sigma, \tag{3.55}$$

where

$$\begin{aligned} \Pi'_d &\equiv \alpha_{46}\alpha_{65}\alpha_{53}\alpha_{31}\alpha_{12}\alpha_{13} \\ \Pi'_h &= \alpha_{46}\alpha_{68}\alpha_{87}\alpha_{75}\alpha_{53}\alpha_{31}\alpha_{12}\alpha_{13}. \end{aligned} \tag{3.56}$$

The nature of the rate constant selections contained in Π'_d and Π'_h can be seen from Fig. 3.3.

If, for brevity, we now designate

$$\begin{aligned} \Pi_{a+}\Sigma_a &= a, \qquad \Pi_{b-}\Sigma_b = b, \qquad \Pi_{c-}\Sigma_c = c, \\ \Pi'_d\Sigma_d &= d, \qquad \Pi_{f-}\Sigma_f = f, \qquad \Pi'_h = h, \end{aligned} \tag{3.57}$$

and use the identity following Eq. 3.14 for $e^{(X_1+X_T)/kT}$, the thermodynamic fluxes in Eq. 3.45 take the form

$$\begin{array}{lcccc} & (e^{X_1/kT}-1) & (e^{X_2/kT}-1) & (e^{X_T/kT}-1) & \begin{array}{c}(e^{X_1/kT}-1)\\ \times\,(e^{X_T/kT}-1)\end{array} \\ \hline J_1^{\infty}\Sigma/N = & b+d+f+h & -d-h & f+h & f+h \\ J_2^{\infty}\Sigma/N = & -d-h & a+d+h & -h & -h \\ J_T^{\infty}\Sigma/N = & f+h & -h & c+f+h & f+h \end{array} \tag{3.58}$$

This should be compared with Eqs. 3.16 and 3.17. The quantities $a, \ldots, h$ are all positive. These are "flux–force" equations (with the coefficients shown as a matrix), but the forces appear in an unconventional way and the equations are not restricted to near-equilibrium steady states. Note that nonlinear force terms are present, and that the coefficients of linear terms obey reciprocal relations. These coefficients are properties of the steady state, not of the equilibrium state. In particular, if X_1, X_2, X_T are varied by changing c_{1A}, c_{2A}, c_T (Eqs. 3.39–3.41), the coefficients a, b, c, f will also vary (because of the Σ_κ—see above). Of course Σ is also a function of c_{1A}, c_{2A}, c_T. Coupling between the three fluxes is evident from the existence of cross-coefficients.

In summary, Eqs. 3.58 give explicit expressions for the observable steady-state fluxes as functions of the observable forces and of all the rate constants of the model. The rate constants are organized according to different cycle contributions. As a consequence of this type of organization, the matrix of coefficients in Eqs. 3.58 is quite informative even by itself—i.e., without explicit display of the individual rate constants.

3.4 Reciprocal Relations and Irreversible Thermodynamics

The systems being examined in this book are in general at steady states very far from equilibrium. The classical irreversible thermodynamics (1) of Onsager applies only to first (linear) departures *from equilibrium*. For example, in the Onsager domain, the J_i^∞ in Eqs. 3.58 would be linear functions of the X_i:

$$\begin{aligned} J_1^\infty &= L_{11} X_1 + L_{12} X_2 + L_{1T} X_T \\ J_2^\infty &= L_{21} X_1 + L_{22} X_2 + L_{2T} X_T \\ J_T^\infty &= L_{T1} X_1 + L_{T2} X_2 + L_{TT} X_T, \end{aligned} \tag{3.59}$$

where the L_{ij} are constants (kinetic properties of the *equilibrium* state) and **L** is a symmetrical matrix (hence, the reciprocal relations $L_{ij} = L_{ji}$). Offhand one would expect, then, that irreversible thermodynamics could be of significance in the present context only in that it represents a nonphysiological "limiting law" type of behavior for any model or experimental system. This behavior is of course of some theoretical and experimental interest, for example, as a self-consistency test (reciprocal relations), but it would have no direct applicability to the actual in vivo steady state.

The above point of view certainly seems appropriate in the problem of muscle contraction (Chapter 5). This follows simply from the fact that, *near equilibrium*, the force-velocity and flux (ATP)-velocity curves would be linear (13, 14) but, experimentally, these curves are far from linear. Thus the linear-departure-from-equilibrium realm of irreversible thermodynamics does not extend to the conditions actually present in a contracting muscle. A rather thorough analytical examination (14) of a particular though arbitrary muscle model, from equilibrium to far-from-equilibrium, is very helpful in understanding this situation. It is interesting, incidentally, in view of Eqs. 3.16, 3.17, and 3.58, that the range of "linear" behavior of this model, away from equilibrium, is extended considerably (14) on using $e^{X_T/kT} - 1$ as an independent variable in series expansions rather than X_T/kT; i.e., on using $(c_T/c_T^e) - 1$ in place of $\ln(c_T/c_T^e)$. See Table 4.9 in this connection.

It seems unlikely, as well, that the linear *near-equilibrium* regime in, say, the Na,K-ATPase active transport system would remain linear as far from equilibrium as, say,

$$\begin{aligned} X_T = \mu_T - \mu_D &\cong 12.5 \quad \text{kcal mol}^{-1} \\ &\cong 21.5kT \end{aligned} \tag{3.60}$$

in view of the fact that we would normally require $X_T \ll kT$ for linearity. Such an extensive linear domain is not impossible but it would be very surprising indeed.

Of course, not all active transport systems work this far from equilibrium, either in vivo or in controlled experiments. Indeed, there are preliminary indications that some less extreme systems do in fact behave linearly (15) under conditions of practical interest. Furthermore, there is no reason to expect that every free energy transducing system would possess a linear domain of the same extent (in units of kT). Hence the question appears to be an open one at the present time (16).

In any case, linear or nonlinear, accurate and extensive flux-force measurements will obviously be of great assistance in the future in the elimination of inadequate molecular models of active transport and other free energy transducing systems. As an aside, however, it might be added that, just as transient (17, 18) rather than steady measurements provide a more severe test of kinetic molecular models of muscle contraction, the same might be expected to be true for other transducing systems (19). For essentially the same reason, noise measurements should also prove to be very valuable (20–24).

It should perhaps be emphasized that, in the preceding discussion, we have been referring strictly to the extent of the linear *near-equilibrium* domain and not to a possible separate steady-state region of linear behavior far from equilibrium (25). Our primary interest in the former case arises from the fact that this is the only linear domain to which the basic Onsager theory applies.

Single Cycle; Metastable Equilibrium

The above discussion applies to the *general* case of a multicycle diagram. In a system with several forces but only a single significant cycle, the near-equilibrium question is somewhat different. As an example, let us take Section 3.3 with cycle h as the only important cycle (Section 3.1, with cycle a only, is another example). In this case there is only one independent operational flux, $J_h = J_1^\infty = -J_2^\infty = J_\mathrm{T}^\infty$, and only one independent operational force, $X_h = X_1 - X_2 + X_\mathrm{T}$. Now even if X_1, X_2, and X_T are all far from zero, we could have the *net* force $X_h = 0$. Necessarily, then (Eq. 3.26), $J_h = 0$. This is a *metastable* rather than a complete ($X_1 = X_2 = X_\mathrm{T} = 0$) equilibrium. When X_h and J_h are nonzero but small, the system is within the range of irreversible thermodynamics (as in Eq. 3.70, below).

Clearly, in this example and for this purpose, the magnitude of X_T itself (as in Eq. 3.60) is not vital. Instead, the magnitude of the *net* force X_h is important.

Reciprocal Relations and Cycles

Before considering the near-equilibrium situation, we pursue further the "reciprocal relations" already mentioned, for an *arbitrary steady state*, in connection with Eqs. 3.16, 3.17, and 3.58. Let us refer particularly to Eqs. 3.58 since that case is somewhat more general. Looking at the *linear terms only* (i.e., those in $e^{X_i/kT} - 1$), we see that the reciprocal relations between the off-diagonal elements are simply consequences of the fact that *individual* cycle fluxes have all linear coefficients of the *same magnitude* and of *proper sign*:

$$\begin{pmatrix} d & -d & 0 \\ -d & d & 0 \\ 0 & 0 & 0 \end{pmatrix}, \quad \begin{pmatrix} f & 0 & f \\ 0 & 0 & 0 \\ f & 0 & f \end{pmatrix}, \quad \begin{pmatrix} h & -h & h \\ -h & h & -h \\ h & -h & h \end{pmatrix} \tag{3.61}$$

These are the linear submatrices contributed by cycles d, f, and h to the total symmetrical matrix of Eqs. 3.58. The total matrix is necessarily symmetrical because each submatrix is. Therefore the reciprocal relations we encounter are fundamentally a property of *individual cycles*, in particular of those cycles that contain two or more thermodynamic forces.

Any cycle κ that contains two or more forces can be handled as in Eqs. 3.51–3.55. That is, we use the rate constant relation found from detailed balance at equilibrium ($\Pi_{\kappa+} = \Pi_{\kappa-}$) to factor out nonvariable rate constants from $\Pi_{\kappa+} - \Pi_{\kappa-}$. The remaining (unfactored) rate constant ratios can then be put in the form

$$[e^{(X_1 + X_2)/kT} - 1] - [e^{(X_3 + X_4)/kT} - 1] \tag{3.62}$$

if, for example, there are two forces (X_1, X_2) in the direction $\kappa+$ and two (X_3, X_4) in the direction $\kappa-$. The generalization of 3.62 to any number of forces in either direction is obvious. Thus we can write

$$J_\kappa = \{[e^{(X_1 + X_2)/kT} - 1] - [e^{(X_3 + X_4)/kT} - 1]\}\kappa \tag{3.63}$$

where κ (a property of the steady-state) collects all other factors that contribute to this cycle flux. If we rewrite the square brackets in Eq. 3.63 (for any number of forces) in terms of the quantities $e^{X_i/kT} - 1$, as, for example, following Eq. 3.14, it is easy to see that there will always be linear terms in $e^{X_i/kT} - 1$ with coefficients $+1$. Thus, separating out these linear terms, Eq. 3.63 becomes

$$J_\kappa = [(e^{X_1/kT} - 1) + (e^{X_2/kT} - 1) - (e^{X_3/kT} - 1) - (e^{X_4/kT} - 1)]\kappa$$
$$+ \text{higher terms.} \tag{3.64}$$

If the term containing X_k here (or in any other example) is $s_k(e^{X_k/kT} - 1)\kappa$ and the X_j term is $s_j(e^{X_j/kT} - 1)$, where $s_k = \pm 1$ and $s_j = \pm 1$, then the contributions of J_κ to J_k^∞ and to J_j^∞ (as in Eqs. 3.45) are necessarily $s_k J_\kappa$ and $s_j J_\kappa$, respectively. The linear X_k and X_j terms in J_k^∞ and J_j^∞ are then

$$\begin{aligned} J_k^\infty &= \kappa(e^{X_k/kT} - 1) + s_k s_j \kappa(e^{X_j/kT} - 1) + \cdots \\ J_j^\infty &= s_j s_k \kappa(e^{X_k/kT} - 1) + \kappa(e^{X_j/kT} - 1) + \cdots, \end{aligned} \tag{3.65}$$

where we have put $s_k^2 = s_j^2 = 1$. Since $s_k s_j \kappa = s_j s_k \kappa$, we see that single cycles will always exhibit reciprocal relations and also that they will have all coefficients of the same magnitude in submatrices such as in 3.61.

Because the coefficients in Eqs. 3.58, say, are not themselves constant when the forces vary, the reciprocal relations just discussed have no particular physical significance at an arbitrary steady state. But they do become fundamental near equilibrium (see below). It is important, though, that the reciprocal relations *originate* as *properties of single cycles* at an arbitrary steady state. This is another dividend that follows upon the introduction of cycle fluxes.

There is one possible complication that has not been mentioned heretofore. Suppose the same force (e.g., T → D) appears, say, twice in the same cycle κ and in the same direction. Where $e^{X_T/kT}$ would appear with one T → D per cycle, we now have $e^{2X_T/kT}$ and if, say, $J_T^\infty = J_\kappa + \cdots$ (cycles involving T) with one T → D, we now have $J_T^\infty = 2J_\kappa + \cdots$. Using

$$e^{2X_T/kT} - 1 = (e^{X_T/kT} - 1)^2 + 2(e^{X_T/kT} - 1),$$

it is easy to verify that reciprocal relations are not upset.

Irreversible Thermodynamics

According to the by now well-established theory due to Onsager (1), the thermodynamic fluxes and forces near equilibrium will be related as, for example, in Eqs. 3.59, where $L_{ij} = L_{ji}$ and furthermore the L_{ij} are constants (properties of the equilibrium state)—i.e., they are not functions of the X_i. Thus Eqs. 3.59 represent the leading terms in true series expansions of the J_i^∞ in powers of the X_i. The range of validity of these linear relationships is a matter for experimental test and presumably varies from case to case, as already mentioned. They certainly apply, though, when all $X_i \ll kT$. At equilibrium, of course, all $X_i = 0$.

It is an important check on the self-consistency of the present formalism to verify that reciprocal relations $L_{ij} = L_{ji}$ do indeed follow from the formalism. This has practically been proved already in the preceding subsection (Eqs. 3.65). All we need, in addition, is to use the expansion

$$e^{X_i/kT} - 1 = (X_i/kT) + \cdots$$

and set all variable concentrations in κ at their equilibrium values, giving κ^e. As an explicit example, Eqs. 3.58 become, near equilibrium,

$$\begin{array}{lccc}
 & X_1/kT & X_2/kT & X_T/kT \\
\hline
J_1^\infty \Sigma^e/N = & b^e + d + f^e + h & -d - h & f^e + h \\
J_2^\infty \Sigma^e/N = & -d - h & a^e + d + h & -h \\
J_T^\infty \Sigma^e/N = & f^e + h & -h & c^e + f^e + h.
\end{array} \qquad (3.66)$$

Similarly, Eqs. 3.16 and 3.17 become

$$J_1^\infty \Sigma^e/N = (a+b)(X_1/kT) + a(X_2/kT)$$
$$J_2^\infty \Sigma^e/N = a(X_1/kT) + (a+2c')(X_2/kT). \tag{3.67}$$

In both of these examples, the reference concentrations ($c_{1B}, c_{2B}, c_{ADP}, c_{P_i}$) are held constant in the approach to equilibrium.

The corresponding verification of a reciprocal relation in muscle contraction models (26) is a rather more complicated matter (Chapter 5).

Simple Enzyme System Near Equilibrium

We return again to the simple enzyme–substrate–product system of Fig. 1.1 and Eq. 3.35. This is a generalization (2, 11) of the well-known Michaelis–Menten scheme. There is only one flux and one force (hence, no reciprocal relation or free energy transfer), but the model is simple enough to show the explicit form for J^∞ near equilibrium. At an arbitrary steady state, Eq. 3.35 can be written

$$J^\infty = N(e^{X/kT} - 1)\alpha_{21}\alpha_{32}\alpha_{13}/\Sigma \tag{3.68}$$

where, using Eq. 3.33 for α_{12},

$$\Sigma = \alpha_{32}\alpha_{21} + \alpha_{23}\alpha_{31} + \alpha_{21}\alpha_{31} + \alpha_{13}(\alpha_{32} + \alpha_{23} + \alpha_{21}) + \alpha_{12}^e\, e^{X/kT}\,(\alpha_{32} + \alpha_{31} + \alpha_{23}). \tag{3.69}$$

Here again we are assuming that c_S is variable and c_P constant. Near equilibrium, $J^\infty = LX$ (a linear flux-force relation), where

$$L = N\alpha_{21}\alpha_{32}\alpha_{13}/\Sigma^e kT \tag{3.70}$$

and Σ^e is obtained from Eq. 3.69 on putting $X = 0$. L is a *kinetic* property of the system, *at equilibrium*. In this case we could easily derive explicit higher terms in J^∞ expressed as a power series in X (with all coefficients properties of the equilibrium state). Clearly, for this simple but fundamental system, the near-equilibrium linear regime is valid only when $X \ll kT$. This condition is seldom met in real biochemical examples. When it is not met, one must return to Eq. 3.68, which applies to steady states arbitrarily far from equilibrium.

REFERENCES

1. T. L. Hill, "Thermodynamics for Chemists and Biologists," Chapter 7. Addison-Wesley, Reading, Massachusetts, 1968.
2. T. L. Hill and O. Kedem, *J. Theoret. Biol.* **10**, 399 (1966).
3. T. L. Hill, *Progr. Biophys. Mol. Biol.* **29**, 105 (1975).

4. T. L. Hill, E. Eisenberg, Y. Chen, and R. J. Podolsky, *Biophys. J.* **15**, 335 (1975).
5. T. L. Hill, *J. Theoret. Biol.* **10**, 442 (1966).
6. D. Oesterhelt and W. Stoeckenius, *Proc. Nat. Acad. Sci. U.S.* **70**, 2853 (1973).
7. R. Henderson and P. N. T. Unwin, *Nature* (*London*) **257**, 28 (1975).
8. T. L. Hill and R. M. Simmons, *Proc. Nat. Acad. Sci. U.S.* **73**, 95 (1976).
9. T. L. Hill and R. M. Simmons, *Proc. Nat. Acad. Sci. U.S.* **73**, 336, 2165 (1976).
10. J. Keizer, *J. Theoret. Biol.* **49**, 323 (1975).
11. H. R. Mahler and E. H. Cordes, "Biological Chemistry," 2nd ed., p. 282, Harper, New York, 1971.
12. R. Whittam and A. R. Chipperfield, *Biochim. Biophys. Acta* **415**, 149 (1975).
13. T. L. Hill, *Progr. Biophys. Mol. Biol.* **28**, 267 (1974).
14. T. L. Hill and Y. Chen, *Proc. Nat. Acad. Sci. U.S.* **71**, 2478 (1974).
15. A. Essig, *Biophys. J.* **15**, 651 (1975).
16. S. R. Caplan, *Current Topics Bioenerg.* **4**, 2 (1971).
17. M. M. Civan and R. J. Podolsky, *J. Physiol.* **184**, 511 (1966).
18. A. F. Huxley and R. M. Simmons, *Nature* (*London*) **233**, 533 (1971).
19. B. Chance and G. R. Williams, *Advan. Enzymol.* **17**, 65 (1956).
20. J. R. Segal, *Biophys. J.* **12**, 1371 (1972).
21. Y. Chen and T. L. Hill, *Biophys. J.* **13**, 1276 (1973).
22. Y. Chen, *J. Chem. Phys.* **59**, 5810 (1973).
23. Y. Chen, *J. Theoret. Biol.* **55**, 229 (1975).
24. Y. Chen, *Proc. Nat. Acad. Sci. U.S.* **72**, 3807 (1975).
25. H. Rottenberg, *Biophys. J.* **13**, 503 (1973).
26. Y. Chen, *J. Theoret. Biol.* **51**, 419 (1975).

Chapter 4 Free Energy Levels of Macromolecular States

In the first three chapters we examined connections between the rate constants of a diagram and kinetic properties on a larger scale or higher level: cycle fluxes and forces; and thermodynamic fluxes and forces. Here we pursue another rate constant connection, but this time on the level of the individual transitions themselves: the rate constants of a diagram are directly tied to the relative free energy levels of the different macromolecular states that are participating in the diagram (1–5). A study of the free energies of the individual states provides a more detailed understanding of the "drive" that operates in these cycling, steady-state, free-energy-transducing systems and also of the origin of the free energy dissipation.

As already mentioned in Chapter 1, some of the methodology in the first three chapters is not needed or becomes relatively simple if the diagram consists of a single cycle only (see also Appendix 4). This is much less true of the present chapter and of the following chapter (on muscle contraction) because primary attention is necessarily focused here on properties of states rather than of cycles.

4.1 Free Energy Levels of the States

The remarks at the beginning of Sections 1.1 and 1.2, on the nature of the systems being considered, should be reviewed. A few additional comments in the same vein may be appropriate here, before turning to the actual subject of this section.

The N systems (macromolecular units) of the ensemble are *equivalent* and *independent* of each other. If we are dealing, say, with an enzyme complex comprised of interacting enzymes, the entire complex is the "system" or "unit" (Chapter 7). The systems may be free solute molecules in a solution, or they may be immobilized on a surface, in a membrane, etc. In either case, their independence makes it possible to assign thermodynamic functions, in particular, relative free energy values, to the different possible discrete states of an *individual* system or unit. The strict justification for this, from a purely thermodynamic point of view, comes from the thermodynamics of small systems (6, 7). In statistical mechanics, there is essentially no problem since statistical mechanics handles small systems in the same way that it deals with large systems (6–8). In many cases, a potential of mean force (8) would have to be introduced to take care of interactions with the surroundings (e.g., the solvent).

The assignment of a free energy level to a single system in a given state of a kinetic diagram implies that, when in this state, the system may be regarded as an equilibrium system. This is an important point that perhaps should be expanded upon (1).

Diagrams and cycles are of course commonplace in biochemistry. Whenever a diagram (states, rate constants) is written down in an attempt to represent experimental data, the following is implied: the given rate constants set a certain time scale for the diagram and associated kinetics (e.g., millisecond, or second, or microsecond, etc.); transitions between states occur "instantaneously" (on the time scale of the diagram); each state shown in the diagram is actually a composite of (in general) a very large number of more elementary "internal" substates; within any state, transitions between its substates are necessarily very fast compared to the transitions shown in the diagram (for otherwise, these substates, or groups of these substates, should themselves be included as separate states in an expanded diagram); since, for any state, intersubstate transitions are very fast on the time scale of the diagram, the substates of each state maintain internal equilibrium with each other as the states themselves follow the comparatively slow and usually nonequilibrium kinetics prescribed by the diagram. Thus, in general, the states are not in equilibrium *with each other*, yet each state is an equilibrium state *internally*. If, because of relatively slow internal transitions, a certain state is *not* in internal equilibrium, it must be subdivided (thus expanding the diagram) until each newly defined state *is* in internal equilibrium (to the required order of approximation).

Hence, essentially by definition, the states of a diagram may be regarded as equilibrium states, *internally*. We emphasize again that these conclusions are implicit whenever a biochemical kinetic scheme is set down—though usually they are not spelled out.

Explicit use of free energy levels, as in the present book, was made first in papers on muscle contraction (1, 9) where it was found very fruitful to relate the free energy levels of the different biochemical states of a myosin cross-bridge to the first-order rate constants governing transitions between these states and to the details of the conversion of ATP free energy into mechanical work. These ideas were then extended to other problems and in other ways, in subsequent papers by Hill (2), Hill and Simmons (3, 4), and Simmons and Hill (5). These latter papers are summarized and extended somewhat in the present chapter. See also Appendix 6.

Although we are still primarily interested in steady states, transients will be mentioned occasionally in this and the following chapter. The transients are, of course, determined mathematically by equations like Eq. 1.1 (where steady state corresponds to all $dN_i/dt = 0$).

We shall define three kinds of free energy change for any given transition ij: standard, basic, and gross. These will also be referred to as free energy level differences between states i and j, and in fact we shall refer for convenience to the free energy levels of the individual states (i and j) though of course only the level *differences* can have any real operational significance.

Immobilized Macromolecular Systems

Although, as we shall see, the same notation for free energy levels can be used in both cases, we begin by studying an ensemble of immobilized macromolecular systems (say in a membrane) and turn in the next subsection to macromolecules in solution.

Of the N systems in the ensemble, let N_i be the number in state i and define the probability or fraction $p_i = N_i/N$. Consider the ensemble at some particular composition at an arbitrary time t. From the point of view of statistical mechanics, this is an ideal solid solution [see Hill (8), p. 373] with components $i = 1, 2, \ldots, n$ and mole fractions p_i. Let $A_i = -kT \ln Q_i$ be the Helmholtz free energy of a single system in state i [see also Hill (1), Section IIA]. Then the canonical partition function Q of the entire solid solution (ensemble) is

$$Q = N! \prod_i Q_i^{N_i} \Big/ \prod_i N_i!, \tag{4.1}$$

where $N = \sum_i N_i$. The chemical potential of an arbitrary component j in the ensemble is then

$$\mu_j = -kT\, \partial \ln Q/\partial N_j = A_j + kT \ln p_j. \tag{4.2}$$

In taking the derivative here, we have substituted $N = \sum_i N_i$ and held all N_i other than N_j constant. Note that if $p_j = 1$ (all systems in state j), $\mu_j = A_j$. Thus, A_j may be regarded as either the free energy of a single isolated system in state j ($A_j = -kT \ln Q_j$), or the free energy per system if the whole ensemble is in state j ($p_j = 1$). From the form of Eq. 4.2, A_j is also the standard free energy of state j with standard state chosen as $p_j = 1$. Incidentally, we shall use throughout the book the approximation G_j (Gibbs free energy) $\cong A_j$ since the pV_j term is negligible (these are condensed systems). See Hill (1), p. 272, and the Preface of this book, in this connection.

Whereas A_j in Eq. 4.2 is an intrinsic equilibrium property of state j alone, the chemical potential μ_j is obviously a property of the whole ensemble since the population p_j at an arbitrary time t depends in general on *all* the rate constants of the diagram and on boundary conditions (i.e., on the initial state—at $t = 0$—of the ensemble and on any restraints imposed in the time evolution of the ensemble).

The standard free energies A_i of the states i of a single macromolecular system are fundamental in that they can be related to the first-order rate constants (of the diagram) which determine the kinetics of the ensemble. To see this, we have to examine two cases. Consider first the inverse transitions between two states i and j such that no ligand is bound or released in the transitions (e.g., a conformational change E $\rightleftarrows$ E* or ES $\rightleftarrows$ EP). That is, these are *isomeric* transitions. As usual, the first-order rate constants for $i \to j$ and $j \to i$ are designated α_{ij} and α_{ji}, respectively. A hypothetical equilibrium between the two states (with all other transitions in the diagram imagined blocked) can be used to establish the connection between the A's and α's:

$$\alpha_{ij} p_i^e = \alpha_{ji} p_j^e \quad \text{(detailed balance)} \tag{4.3}$$

$$\mu_i = A_i + kT \ln p_i^e = \mu_j = A_j + kT \ln p_j^e \tag{4.4}$$

$$\alpha_{ij}/\alpha_{ji} = \exp[-(A_j - A_i)/kT] \equiv K_{ij}, \tag{4.5}$$

where K_{ij}, defined here, is a dimensionless equilibrium constant. Although $\alpha_{ij}/\alpha_{ji} = p_j^e/p_i^e$ holds only at equilibrium, Eq. 4.5 is valid regardless of the state of the ensemble, that is, even after the imaginary blocks on other transitions of the diagram are removed. This follows because the A's and α's are *intrinsic properties* of states i and j of *each system* that can have nothing to do with what is transpiring in the rest of the ensemble. We have merely used the above hypothetical equilibrium condition of the ensemble, as a convenience, to establish a relation that is *independent* of the state of the ensemble.

The second possibility we have to examine is that a ligand L (or S, or P, etc.) is bound in the transition $i \to j$. In this case, if we consider the equilibrium between states i and j in the presence of L in solution at its *actual* concentration c_L, we have

$$\alpha_{ij}^* c_L p_i^e = \alpha_{ji} p_j^e \quad \text{(detailed balance)}$$

or (4.6)

$$\alpha_{ij} p_i^e = \alpha_{ji} p_j^e,$$

where the (pseudo) first-order rate constant $\alpha_{ij} \equiv \alpha_{ij}^* c_L$. The equilibrium condition is now

$$\mu_i + \mu_L(c_L) = A_i + kT \ln p_i^e + \mu_L = \mu_j = A_j + kT \ln p_j^e. \tag{4.7}$$

Therefore

$$\frac{\alpha_{ij}}{\alpha_{ji}} = \exp\left\{-\frac{[A_j - (A_i + \mu_L)]}{kT}\right\} \equiv K_{ij}, \tag{4.8}$$

where again K_{ij}, defined here, is a dimensionless equilibrium constant. Thus, in binding transitions, μ_L (or μ_S, etc.) must be included in the free energy difference to establish the correct relationship to the first-order rate constant ratio α_{ij}/α_{ji}. Equation 4.8 is the analogue of Eq. 4.5. We see that by "correcting" A_i with $\mu_L(c_L)$, both types of transition can be put on the same footing (1, 2): a free energy change between states is related to the quotient of first-order rate constants.

Alternatively, if we use Eq. 4.6a rather than Eq. 4.6b, and introduce

$$\mu_L = \mu_L^0 + kT \ln c_L \tag{4.9}$$

into Eq. 4.7, we arrive at the more conventional result

$$\frac{\alpha_{ij}^*}{\alpha_{ji}} = \exp\left\{-\frac{[A_j - (A_i + \mu_L^0)]}{kT}\right\} = K_{ij}^*. \tag{4.10}$$

This "second-order" equilibrium constant can also be written

$$K_{ij}^* = p_j^e / c_L p_i^e = K_{ij}/c_L. \tag{4.11}$$

Equation 4.10 involves the standard free energy (chemical potential) μ_L^0 of L whereas Eq. 4.8 contains the actual chemical potential μ_L at c_L, as given by Eq. 4.9. Correspondingly, K_{ij}^* is invariant while K_{ij} depends on the value of c_L.

The kind of free energy change between states i and j that appears in Eqs. 4.5 and 4.8, which is directly related to *first-order* rate constants, will be referred to (3, 4) as a *basic free energy* change. If a ligand L at c_L is involved in the process, $\mu_L(c_L)$ must be included as illustrated in Eq. 4.8. The term "basic" is appropriate because: (a) these free energy changes are time-independent intrinsic properties of each individual system of the ensemble; (b) the use of first-order rate constants throughout the diagram puts all transitions on an equivalent *kinetic* and mathematical basis; and (c) the "correction" with μ_L (or μ_S, etc.), as needed, in the basic free energy change correspondingly puts all of the states of the diagram on an equivalent *thermodynamic* basis so that all transitions become pseudoisomeric. The "correction" with μ_L (instead of μ_L^0) is the thermodynamic equivalent of the "correction" of α_{ij}^* to give a pseudo-first-order rate constant α_{ij}. This uniform first-order, isomeric formalism is the simplest possible.

Ordinary biochemical thermodynamics focuses attention on free energy changes involving substrates, products, ligands, etc. Here the point of view is just the opposite. In order to relate in the simplest way possible to first-order diagram kinetics, we deal with pseudoisomeric *macromolecular* states, free energy levels, and transitions in which substrates, etc., are relegated to a strictly secondary and implicit role. Of course the substrates, etc., come to the fore at the complete cycle level because they then account for the only changes that have taken place as a result of a cycle completion.

The *standard free energy* changes corresponding to the above basic free energy changes are, of course, more familiar and appear in Eqs. 4.5 and 4.10. Note that, for an isomeric process (Eq. 4.5), the standard and basic changes are the same. Standard free energy changes are not particularly useful in the context of the present book because they do not refer to the true concentrations of ligands, substrates, etc., in the actual *kinetic* processes being considered but rather to reference concentrations $c_L = 1$ M, etc. Standard free energies are of course relevant for most equilibrium purposes (10).

A third free energy change of importance relates to the ensemble as a whole, at an arbitrary time t, whereas basic and standard free energy changes are "private" properties of individual systems (and therefore do not depend at all on boundary conditions or the time t). If the ensemble at time t has a composition specified by $p_1, p_2, \ldots, p_n$, then the free energy change *in the ensemble* plus surrounding bath or baths, associated with the processes $i \to j$ in Eqs. 4.5 and 4.8, are

$$\begin{aligned} \mu_j - \mu_i &= A_j - A_i + kT \ln(p_j/p_i) \\ \mu_j - (\mu_i + \mu_L) &= A_j - (A_i + \mu_L) + kT \ln(p_j/p_i). \end{aligned} \tag{4.12}$$

Since these changes refer to the whole ensemble, they are designated *gross free energy* changes (3–5).

For convenience, for all three kinds of free energy changes, we shall refer as well to *free energy levels* of the separate *states*. For example, A_j and A_i in Eq. 4.5 and A_j and $A_i + \mu_L$ in Eq. 4.8 are the "basic free energy levels" of states j and i. Actually, only the differences are significant but the assignment of separate levels to the states has value as a conceptual aid. Note that, in the second example here, we could just as well have used (1) $A_j - \mu_L$ and A_i.

The gross free energy changes in Eqs. 4.12 are, of course, just the conventional Gibbs free energy changes that determine equilibrium in the ensemble (plus baths) or the direction of spontaneous "flow" toward equilibrium. A new name ("gross") is adopted for convenience (to contrast with "basic") but one is also needed because it refers to the relative free energy *levels* of the *macromolecular* states on a *pseudoisomeric* basis. As explained above, this is not a conventional point of view; therefore, some new designation is appropriate.

Although we shall speak, for brevity, of the free energy levels of the *macromolecular* states, it must be kept in mind that the corresponding free energy changes relate to macromolecules *plus* surrounding bath or baths.

From an operational point of view, in theoretical work, basic free energy levels are introduced ab initio as fixed parameters of a model, while the gross free energy levels emerge as calculated macroscopic properties of the ensemble that depend on the p_i and therefore on the boundary conditions and on the time t, in general. The gross levels are of course all equal when there is (pseudoisomeric) equilibrium among all macromolecular states i, but not otherwise.

In its stochastic behavior (Chapter 6), any *individual* system of the ensemble is governed entirely by the α's of the diagram (which are related to the basic free energy levels). The individual system has no knowledge of *ensemble* properties such as the p's, the gross free energy levels, transient versus steady state, etc.—or even whether there *is* an ensemble ($N > 1$). Ensemble properties depend on the *statistics* of these individually completely uncontrolled stochastic systems. Of course the choice of rate constants, the α's, for the individual systems will determine whether, at $t = \infty$, the ensemble statistics will correspond to equilibrium or to a nonequilibrium steady state.

The above remarks need to be modified somewhat in the muscle contraction problem (1) (Chapter 5). See also Appendix 6.

It is convenient to introduce special notation for basic and gross free energy changes and levels. For the process $i \to j$, whether it be an isomeric change or involves binding or release of a ligand, etc., we use

$$\Delta A'_{ij} = A'_i - A'_j, \qquad \Delta\mu'_{ij} = \mu'_i - \mu'_j \tag{4.13}$$

to designate the basic and gross free energy changes, respectively. The quantities on the right represent the basic and gross free energy levels of the states. The primes indicate, on both sides of the equation, that μ_L (or μ_S, etc.) is included *where needed*—i.e., that macromolecular free energies are "corrected," as appropriate, by ligand, substrate, etc., chemical potentials in order to put all states in the diagram on the same (isomeric) footing.

Note that Δ is used in Eqs. 4.13 (and below) in an unconventional way: $\Delta \equiv$ initial–final, rather than vice versa. This is done because, in most steady-state applications, we shall be interested in cases dominated by downhill (in free energy) transitions. Hence *positive* values of $\Delta A'_{ij}$ and $\Delta\mu'_{ij}$ will predominate.

Table 4.1 makes Eqs. 4.13 quite explicit for cases in which $i \to j$ is isomeric, binds L to i to form j, or releases L from i to form j. The parentheses establish the convention for the free energy levels of the separate states.

TABLE 4.1

FREE ENERGY NOTATION FOR PROCESS $i \to j$

Free energy change	Isomeric	Binding	Release
Basic, $\Delta A'_{ij}$	$A_i - A_j$	$(A_i + \mu_L) - A_j$	$A_i - (A_j + \mu_L)$
Gross, $\Delta \mu'_{ij}$	$\mu_i - \mu_j$	$(\mu_i + \mu_L) - \mu_j$	$\mu_i - (\mu_j + \mu_L)$
Standard	$A_i - A_j$	$(A_i + \mu_L^0) - A_j$	$A_i - (A_j + \mu_L^0)$

Table 4.1 also makes it clear that standard → basic requires introduction of the actual c_L of the ligand in solution while basic → gross is accomplished by introduction of the actual p_i and p_j in the ensemble of macromolecular units.

With these definitions of $\Delta A'_{ij}$ and $\Delta \mu'_{ij}$, we now have the important relations, for any change in state $i \to j$,

$$\alpha_{ij}/\alpha_{ji} = \exp(\Delta A'_{ij}/kT) = K_{ij} \tag{4.14}$$

and

$$\Delta \mu'_{ij}(t) = \Delta A'_{ij} + kT \ln[p_i(t)/p_j(t)]. \tag{4.15}$$

If states i and j are in equilibrium with each other, $\Delta \mu_{ij}^{\prime e} = 0$.

Appendix 6 expands on some of the topics covered so far in this subsection, in a simple special case.

The product of Eq. 4.14 around any cycle κ, say with states numbered 1, 2, ..., m (in counterclockwise order), is (Eq. 3.26)

$$\Pi_{\kappa+}/\Pi_{\kappa-} = e^{X_\kappa/kT} = K_{12}K_{23} \cdots K_{m1}. \tag{4.16}$$

Also, the total thermodynamic force in the cycle, X_κ, is equal to the sum of the successive basic free energy changes around the cycle:

$$X_\kappa = \Delta A'_{12} + \Delta A'_{23} + \cdots + \Delta A'_{m1}. \tag{4.17}$$

If X_κ is positive, not all of the $\Delta A'_{ij}$ here need be positive. We also have

$$X_\kappa = \Delta \mu'_{12}(t) + \Delta \mu'_{23}(t) + \cdots + \Delta \mu'_{m1}(t) \tag{4.18}$$

since the probability terms cancel if Eq. 4.15 is summed around the cycle (at arbitrary t). X_κ is, of course, independent of t because it depends only on fixed concentrations of ligands, substrates, etc. We shall see several examples of Eqs. 4.17 and 4.18 below.

The overall "drive" around the cycle κ is provided by X_κ. This is both an operational thermodynamic force, or sum of such forces, and a function of the cycle rate constants, as given by Eq. 4.16. This of course implies that, in making up any model, the rate constants of each cycle must be selected to be consistent with the thermodynamic forces in the cycle.

Equation 4.17 shows how the overall drive or force is subdivided among the separate transitions of the cycle according to the $\Delta A'_{ij}$ (or α_{ij}/α_{ji}) values. Thus, $\Delta A'_{ij}$ is the "drive" associated with the transition ij. This subdivision again emphasizes that, in general, no transition or "step" can be singled out as completely controlling the steady-state cyclic action. All steps are involved, one way or another. By the same token, if the cycle contains two or more forces so that free energy transduction is possible (as, for example, in Sections 3.1 and 3.3), the transduction is an indivisible property of the whole cycle (Eq. 4.17) and cannot be assigned to any one transition or $\Delta A'_{ij}$. In any case, the kinetic behavior of the ensemble (e.g., at steady state) is by no means completely determined by the $\Delta A'_{ij}$ values (see Sections 4.3 and 4.4).

The subdivision in Eq. 4.17 is an invariant ("basic") property (of any given model) that belongs to each *individual* system in the ensemble. On the other hand, the subdivision in Eq. 4.18 is a time-dependent property of the ensemble taken as a whole.

Incidentally, it is interesting that the relation between individual transition rate constant ratios and basic free energy differences, as given by Eq. 4.14, carries over to the cycle level. Thus $e^{X_\kappa/kT}$ in Eq. 4.16 is also equal to $k_{\kappa+}/k_{\kappa-}$ (Eq. 2.14). The analogy is not perfect, however, since a change of biochemical state is involved in Eq. 4.14 but completion of a cycle (Eq. 4.16) leaves the system unchanged (except for the transport, etc., accomplished).

There are two alternative ways to subdivide the overall drive in Eq. 4.17, though these lack any thermodynamic significance. The subdivision in Eq. 4.17 is by *transition pair*: the "drive" associated with the transition pair ij is $kT \ln(\alpha_{ij}/\alpha_{ji})$. Perhaps more interesting kinetically is a subdivision by *state*: the drive associated with state j, in the sequence ijk, is $kT \ln(\alpha_{jk}/\alpha_{ji})$. This quantity summed around the cycle κ also gives X_κ (Eq. 4.17). The ratio α_{jk}/α_{ji} gives the relative probability of a *forward* (jk) transition, out of state j, compared to a *backward* transition (ji). We can subdivide (in either the "transition" or the "state" system above) even further and assign a "drive" $kT \ln \alpha_{ij}$ to each forward transition and $-kT \ln \alpha_{ji}$ to each backward transition. These also add to X_κ (around the cycle). These last "free energy changes" obviously depend on the units used for the α's (sec^{-1}, msec^{-1}, etc.).

Finally, in this subsection, we mention three complications:

(a) Myosin in muscle is an example of an "immobilized" enzyme. In the muscle problem (Chapter 5), a positional variable x must be introduced (1, 11) that locates the nearest actin attachment site relative to the myosin cross-bridge in question. From a theoretical point of view, it is this feature that leads to the introduction of mechanical force and work into the analysis and hence that makes possible an understanding of how ATP free energy is converted into mechanical work (1). Some of the A_i and some or all of the α_{ij} (depending on the model) are functions of x (Chapter 5). The above free energy considerations still apply but, in effect, we have to consider a different ensemble of myosin molecules in each interval $x, x + dx$. This generalization is, however, not needed in studies of heavy meromyosin (or S1) + ATP + F-actin in solution since there is no variable x in this case.

(b) If we are concerned with charged macromolecular systems in a membrane, a membrane potential, and charged ligands, then electrostatic potential terms must be added to the nonelectrostatic free energy terms. For a ligand with charge $z\varepsilon$ (e.g., $z = +1$ for K^+) in a bath at potential ψ, μ_L^0 is replaced by $\mu_L^0 + z\varepsilon\psi$ in Eq. 4.9 for μ_L, and in Table 4.1. Also, each macromolecular free energy A_i will now include a similar term for each of its charges (including bound ligand) but using the local ψ at each charge. To be more precise, aside from charged ligands, we need include only those macromolecular charges that change position in the membrane electrostatic potential field as a result of transitions included in the diagram. These adjustments are vital in particular models but they do not alter the general formalism except in trivial ways. It is for this reason that our text examples exclude electrostatic effects.

Appendix 3 contains an explicit example.

(c) What becomes of basic and gross free energy changes when the diagram is reduced by eliminating a state (see Appendix 1)? If state B, in the sequence of states A, B, C (Fig. A1.1), is dropped from the diagram because it is a low-probability transient intermediate, then for the "new" overall reaction A → C we have

$$\Delta A'_{AC} = \Delta A'_{AB} + \Delta A'_{BC} \tag{4.19}$$

and

$$\begin{aligned}\Delta\mu'_{AC} &= \Delta\mu'_{AB} + \Delta\mu'_{BC} \\ &= \Delta A'_{AC} + kT\ln(p_A/p_C).\end{aligned} \tag{4.20}$$

That is, the consecutive free energy changes are simply added in this case and p_B disappears by cancellation. Equation 4.19 is consistent with (see Eq. 4.14)

$$k_+/k_- = (k_1/k_{-1})(k_2/k_{-2}) \tag{4.21}$$

from Eq. A1.6 (Appendix 1).

If B and C in the sequence of states A, B, C, D are in a relatively fast equilibrium with each other (Fig. A1.3), there is, in effect, a single intermediate S = B + C in the "new" sequence A, S, C. In this case the basic free energy level of S is given by [see Hill (8), p. 183]

$$e^{-A_S'/kT} = e^{-A_B'/kT} + e^{-A_C'/kT} \tag{4.22}$$

and the gross free energy level by

$$\mu_S' = \mu_B' = \mu_C' . \tag{4.23}$$

It is easy to verify that Eq. 4.22 is consistent with Eqs. A1.22 and A1.25.

Macromolecular Systems in Solution

We have gone into considerable detail above, for the immobilized case alone, because the solution case can be incorporated into this discussion without any difficulty or significant change.

In the solution case, we have a multicomponent mixture of macromolecules, in the various states i, that may be regarded as solutes in a mixed solvent (water, salt, ligands, etc.). The solution is necessarily dilute with respect to macromolecules since they are assumed to be kinetically independent. The chemical potential of those systems in state i can then be written as

$$\mu_i = \mu_i^0 + kT \ln c_i , \tag{4.24}$$

where c_i is the molar concentration in solution. This is the analogue of Eq. 4.2. In order to be able to use the notation in Eq. 4.2 for both cases (immobile and free), let us introduce $c_i = cp_i$ in Eq. 4.24, where $c = N/V$, the *total* macromolecular concentration, is the same *constant* for all states i, and p_i is the fraction of macromolecular systems in state i. Thus the equations of the preceding subsection apply to the present situation as well, if we understand A_i to mean here

$$A_i \equiv \mu_i^0 + kT \ln c. \tag{4.25}$$

Since only free energy differences, such as $A_i - A_j$, have physical significance in any case, introduction of the term $kT \ln c$ is a mere formality. Use of μ_i^0 as essentially the free energy A_i of a single macromolecule (in state i) in solution is equivalent to a procedure previously introduced in the thermodynamics of small systems (6, pp. 50–58).

Incidentally, charged ligands and macromolecules require no special consideration in the solution case (whereas they do in membrane problems—see above). The electrostatic potential in the single solution is chosen as $\psi = 0$. There will be a nonoperational local potential, a function of position, within each macromolecule (12). But electrostatic contributions resulting therefrom (see Appendix 3) would be included implicitly in the A_i.

Direction of Spontaneous Transition

Most of the remainder of this chapter will be based on special cases, as examples, but in this subsection we deal with one property that can easily be discussed in a completely general way. Consider an ensemble of N systems (immobilized or in solution) with an arbitrary kinetic diagram. At an arbitrary time t (this need not be a steady state), let p_i be the probability of state i. For any transition ij in the diagram, the net mean transition flux $i \to j$ is (Eq. 1.15)

$$J_{ij}(t) = N[\alpha_{ij} p_i(t) - \alpha_{ji} p_j(t)]. \tag{4.26}$$

Also, Eqs. 4.14 and 4.15 can be combined to give

$$\alpha_{ij} p_i(t)/\alpha_{ji} p_j(t) = e^{\Delta\mu_{ij}'(t)/kT}. \tag{4.27}$$

On comparing Eqs. 4.26 and 4.27, we see that the transition flux J_{ij} for any transition ij *always has the same sign* as the gross free energy level difference $\Delta\mu'_{ij}$ (at any time t). For example, if i has the higher gross free energy level, the net mean flux will be in the direction $i \to j$ (there are, of course, stochastic exceptions in single systems or in small groups of systems). This is just the second law of thermodynamics at work on individual reactions of a complex reaction scheme. Keizer (13) has discussed this topic in a more general context.

Thus, net reaction (positive flux) always occurs in a *downhill* direction with reference to a set of gross free energy levels. This is *not* true of the invariant basic free energy levels. This point will be amply illustrated in the examples below.

Rate of Entropy Production

Each event $i \to j$ in a single system contributes a free energy drop $\Delta\mu'_{ij}$ (Eqs. 4.12) to the whole ensemble (plus baths). Hence $J_{ij}\, \Delta\mu'_{ij}$ is the net rate of free energy dissipation (entropy production) contributed by the process $i \to j$. We have just seen that the product $J_{ij}\, \Delta\mu'_{ij}$ is always ≥ 0, with the equality holding only at equilibrium. The *total* rate of entropy production in ensemble plus baths, at any time t, is therefore

$$T\frac{d_i S}{dt} = \sum_{ij} J_{ij}(t)\, \Delta\mu'_{ij}(t) \geq 0, \tag{4.28}$$

where the sum is over all lines in the diagram (the direction chosen along each line is immaterial because a change in direction reverses both signs in the product).

In muscle, with a different ensemble at each x, since each $J_{ij}(t, x)\, \Delta\mu'_{ij}(t, x) \geq 0$, we not only have a sum ≥ 0 for each x when adding contributions from all transitions as in Eq. 4.28, but we also have an integral ≥ 0 for each transition ij on summing contributions from all intervals x, $x + dx$. The total rate of entropy production is obtained from both operations combined: $\sum_{ij}$ and $\int dx$ (Chapter 5).

Equation 4.28 should be compared with Eqs. 3.22 and 3.25 (steady states only). Also, the discussion following Eqs. 3.26 should be noted. The strong resemblance between the pair of equations 3.26 (for cycles) and the pair 4.26 and 4.27 (for transitions) should also be observed. The latter pair are more "detailed" and also apply at arbitrary t.

We have so far encountered three different expressions for the rate of entropy production or free energy dissipation. A fourth appears in Section 4.4 and in the next chapter.

The remainder of the chapter is devoted to assorted examples, with some treated more extensively than others.

4.2 Single-Cycle Examples

Model for Facilitated Diffusion

In this model we introduce only the basic and standard free energy levels. We consider the simple and well-known model in Fig. 4.1a for the facilitated diffusion of an uncharged ligand (L), between bath A and bath B across a membrane, with the aid of a "carrier" that has a single binding site ($\times$) for a ligand molecule. There are N independent and equivalent carriers in the

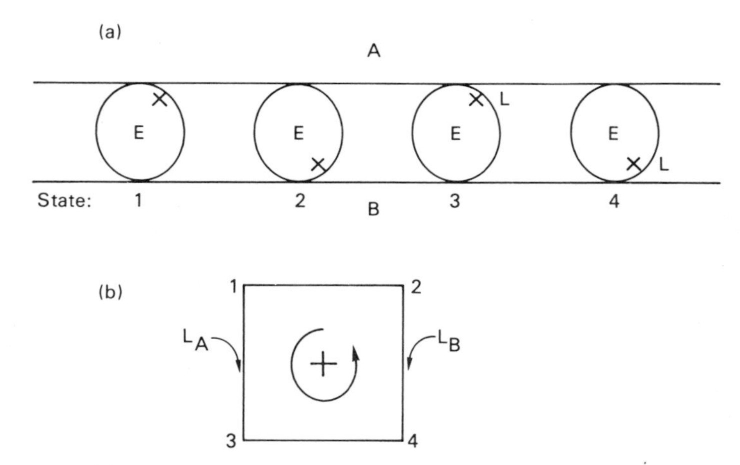

FIG. 4.1. (a) Model for facilitated diffusion, across a membrane, of ligand L between baths A and B. (b) Diagram of model.

membrane sample. Each carrier may exist in four different states, indicated very schematically in Fig. 4.1a. The actual mechanism of translocation across the membrane (conformational change, rotation, etc.) need not be specified. This model is clearly a simplified version of the two-ligand model introduced in Fig. 2.11.

The biochemical diagram is a single cycle in this case, shown in Fig. 4.1b. The rate constants α_{ij} are first-order rate constants for transitions between carrier states with ligand at its actual molar concentrations c_A and c_B in the two baths. That is, $\alpha_{13} = \alpha^* c_A$ and $\alpha_{24} = \alpha^* c_B$.

The condition for equilibrium is $c_A = c_B$, which leads (with $\Pi_+ = \Pi_-$) to the obligatory rate constant relationship $\alpha_{34}\alpha_{42}\alpha_{21} = \alpha_{31}\alpha_{43}\alpha_{12}$. Otherwise ($c_A \neq c_B$) there is a steady state (we do not consider transients) with a net mean flux J^∞ around the cycle (Fig. 4.1b) corresponding to net transport of ligand across the membrane. The expression for J^∞ in the direction A → B is

$$J^\infty = N(e^{X/kT} - 1)\Pi_-/\Sigma, \tag{4.29}$$

where, from Eq. 4.9,

$$X = \mu_A - \mu_B = kT \ln(c_A/c_B) \tag{4.30}$$

and Σ is a sum of $4 \times 4 = 16$ terms since there are four partial diagrams (Fig. 1.10b).

This model has only one force and there is no free energy transduction. The equations of type 4.14 for this model are

$$\alpha_{13}/\alpha_{31} = \exp\{[(A_1 + \mu_A) - A_3]/kT\} = K_{13} \tag{4.31a}$$

$$\alpha_{34}/\alpha_{43} = \exp[(A_3 - A_4)/kT] = K_{34} \tag{4.31b}$$

$$\alpha_{42}/\alpha_{24} = \exp\{[A_4 - (A_2 + \mu_B)]/kT\} = K_{42} \tag{4.31c}$$

$$\alpha_{21}/\alpha_{12} = \exp[(A_2 - A_1)/kT] = K_{21}. \tag{4.31d}$$

On multiplying these together we get

$$\Pi_+/\Pi_- = e^{(\mu_A - \mu_B)/kT} = e^{X/kT} = K_{13}K_{34}K_{42}K_{21}. \tag{4.32}$$

The quantity $\mu_A - \mu_B$ (Eq. 4.30) is the *total* or net actual free energy "drop" (positive or negative) associated with one counterclockwise cycle (a ligand molecule is transported from bath A to bath B). Each of the four members of Eq. 4.32 is equal to unity only at equilibrium ($c_A = c_B$, $\mu_A = \mu_B$). When $\mu_A > \mu_B$, these members are greater than unity and there is a net mean counterclockwise flux (ligand: A → B), given by Eq. 4.29.

For contrast, let us replace the basic free energy changes above by standard free energy changes, as in Eq. 4.10 and Table 4.1. The first and third of Eqs. 4.31 become

$$\begin{aligned} \alpha^*/\alpha_{31} &= \exp\{[(A_1 + \mu^0) - A_3]/kT\} = K_{13}^* \\ \alpha_{42}/\alpha^* &= \exp\{[A_4 - (A_2 + \mu^0)]/kT\} \equiv K_{42}^* . \end{aligned} \tag{4.33}$$

The second and fourth of Eqs. 4.31 are unchanged. Multiplication of the four (modified) equations now yields

$$\alpha_{34}\alpha_{42}\alpha_{21}/\alpha_{31}\alpha_{43}\alpha_{12} = 1 = K_{13}^* K_{34} K_{42}^* K_{21} \tag{4.34}$$

since the total *standard* free energy change for one complete cycle is zero. This also follows from Eq. 4.32 if we put $c_A = c_B = 1$ M (hence, $\mu_A = \mu_B = \mu^0$). These properties obtain whether the actual system is at steady state or at equilibrium (i.e., whatever the actual values of c_A and c_B); hence, they are not very informative.

Figure 4.2a shows a possible illustrative set of basic free energy levels and differences, corresponding to Eqs. 4.31, for the carrier states in Fig. 4.1. In

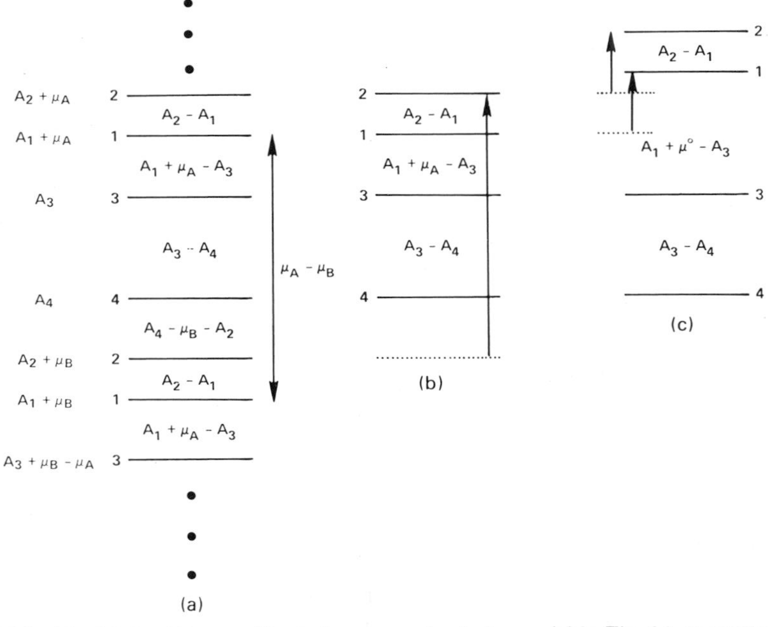

FIG. 4.2 (a) Possible set of basic free energy levels for model in Fig. 4.1, at steady state ($\mu_A > \mu_B$). (b) An equilibrium set of basic free energy levels. See text. (c) A set of standard free energy levels. See text.

this example, we have chosen $\mu_A > \mu_B$ (and $c_A > c_B$) and there is a basic free energy drop in *each* counterclockwise step of the cycle (this need not be the case but the figure is thereby simplified). Each basic free energy difference is related to a first-order rate constant quotient (Eqs. 4.31). Because each step has been assigned a free energy drop, each of these rate constant quotients is greater than unity. The overall basic free energy drop per counterclockwise cycle is $\mu_A - \mu_B$. The set of free energy levels 13421 is *repeated indefinitely* above and below those levels shown in the figure, one set for each cycle.

From a stochastic point of view, at the individual transition level, a carrier performs a biased one-dimensional walk on the free energy levels of Fig. 4.2a. The transition probabilities governing the walk are the first-order rate constants. The bias or "drive" is in favor of downward transitions, as determined by the basic free energy differences (Eqs. 4.31). The net flux ($A \rightarrow B$) contributed by a single carrier is equal to the net number of completed downward cycles (13421). For further details, see Hill (14), Section VI and Chapter 6.

An *equilibrium* special case is shown in Fig. 4.2b. In this case we start with the steady state represented by Fig. 4.2a, hold c_A (and μ_A) constant, but increase c_B (and μ_B) until $c_B = c_A$. The result is the *single* set of equilibrium basic free energy levels in Fig. 4.2b. The dotted level 2 (from Fig. 4.2a) is raised by an amount $\mu_A - \mu_B$ (the three other free energy differences are unchanged) so that now the overall basic free energy change in the cycle 13421 is zero. The quotient α_{42}/α_{24} in Eq. 4.31c is now much less than unity. The relative populations (p_i^e) of the four carrier states now follow a Boltzmann distribution ($p_4^e > p_3^e > p_1^e > p_2^e$), as determined by the three basic free energy differences (Fig. 4.2b). This is *not* true of the steady-state populations (p_i^∞) and levels in Fig. 4.2a (for an explicit example, see Eqs. 4.47–4.51).

Figure 4.2c illustrates a possible set of *standard* free energy levels, related to second-order rate constants as in Eqs. 4.33. The differences $A_2 - A_1$ and $A_3 - A_4$ are unchanged. We suppose in this example that $c_A = c_B < 1$ M in Fig. 4.2b. Hence both levels 1 and 2 in Fig. 4.2b (dotted in Fig. 4.2c) must be raised by the same amount relative to 3 and 4 to arrive at Fig. 4.2c, because the standard state has higher concentrations, $c_A = c_B = 1$ M. There is again only a single set of free energy levels in Fig. 4.2c, since this is an "equilibrium" situation ($c_A = c_B$).

The relative standard free energy levels in Fig. 4.2c and the rate constant ratios in Eqs. 4.31b, 4.31d, and 4.33 are an invariant set (independent of the actual values of c_A and c_B). In contrast, for example, the basic free energy difference between states 4 and 2 and the value of the ratio α_{42}/α_{24} associated with Fig. 4.2a change continuously to values appropriate to Fig. 4.2b as c_B is increased from its initial value (Fig. 4.2a) to its final value $c_B = c_A$ (Fig. 4.2b).

Invariance, of course, always has a certain appeal. But in this case it is a disadvantage. The basic free energy levels (and corresponding first-order rate constants) must be used in order to follow the stochastic time course of the *actual* free energy of any individual system in the ensemble. As cycles are completed by a system, the free energy change $\mu_A - \mu_B$ is pertinent (Eq. 4.17), not $\mu^0 - \mu^0 = 0$.

Simple Enzyme–Substrate–Product Model

We continue here the discussion of the model studied in Section 3.2. The diagram is shown in Fig. 4.3 (see also Fig. 1.1). The enzyme E may be immobilized or in solution. We consider four different subexamples for this same model.

(a) First, we compare basic and standard free energy levels, because there is some contrast here with the preceding subsection.

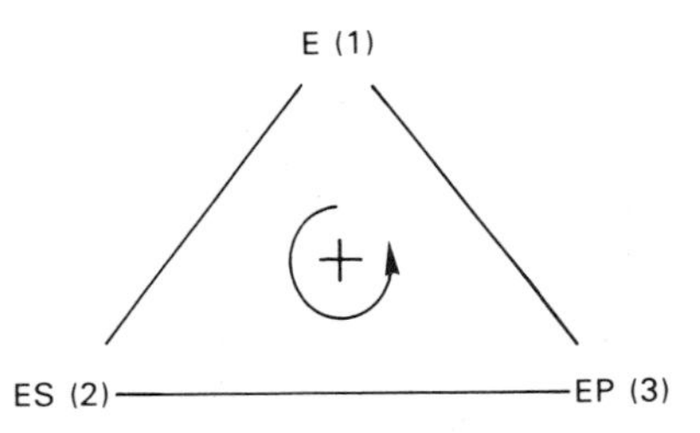

FIG. 4.3 Diagram for enzyme–substrate–product system.

Corresponding to Eqs. 4.31 and 4.32, we have

$$
\begin{aligned}
\alpha_{12}/\alpha_{21} &= \exp\{[(A_1 + \mu_S) - A_2]/kT\} = K_{12} \\
\alpha_{23}/\alpha_{32} &= \exp[(A_2 - A_3)/kT] = K_{23} \\
\alpha_{31}/\alpha_{13} &= \exp\{[A_3 - (A_1 + \mu_P)]/kT\} = K_{31}
\end{aligned}
\tag{4.35}
$$

and

$$\Pi_+/\Pi_- = e^{(\mu_S - \mu_P)/kT} = e^{X/kT} = K_{12}K_{23}K_{31}, \tag{4.36}$$

where μ_S and μ_P are given by Eqs. 3.27. The quantity $\mu_S - \mu_P$ is the actual free energy drop (usually positive) associated with one counterclockwise cycle (one S at c_S is converted to one P at c_P). The equilibrium case is $\mu_S = \mu_P$ (all four members of Eq. 4.36 are equal to unity).

Figure 4.4a shows a hypothetical (infinite) set of basic free energy levels for this system (again each counterclockwise step is chosen arbitrarily to show a basic free energy drop). The first-order rate constants of any model of

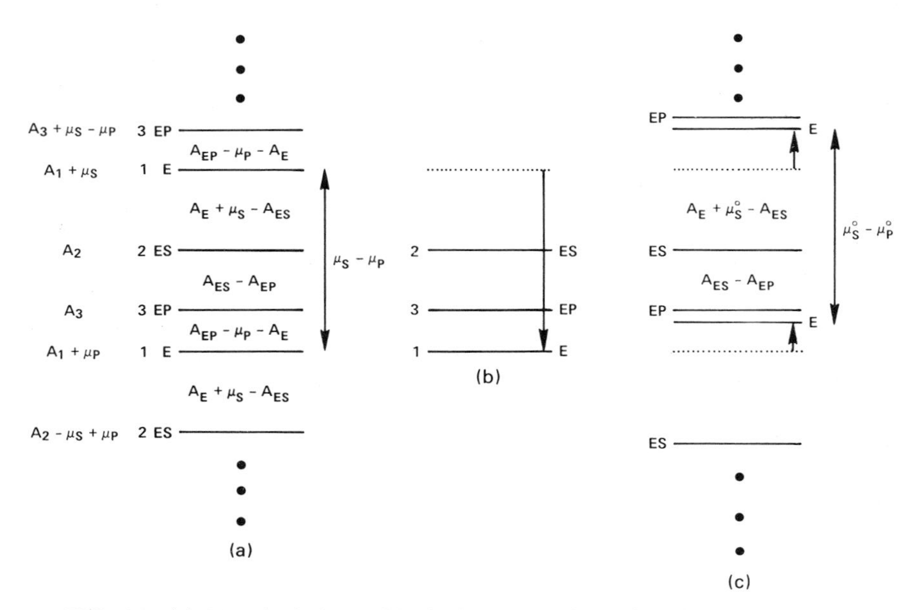

FIG. 4.4 (a) Hypothetical set of basic free energy levels for Fig. 4.3, at steady state ($\mu_S > \mu_P$). (b) An equilibrium set of basic free energy levels ($\mu_S = \mu_P$). See text. (c) Standard free energy levels. See text.

this type imply a set of basic free energy levels, as in Eqs. 4.35 and Fig. 4.4a. Equation 4.36 imposes a restraint on the rate constants chosen for the model since $\mu_S - \mu_P$ is generally a known quantity.

Figure 4.4b shows an equilibrium case. Here we start with Fig. 4.4a, hold c_P and μ_P constant, and decrease c_S and μ_S until $\mu_S = \mu_P$. The result is a single set of three equilibrium basic free energy levels. Correspondingly, $p_1^e > p_3^e > p_2^e$ (Boltzmann distribution).

With standard free energies μ_S^0 and μ_P^0, we have

$$\begin{aligned} \alpha_{12}^*/\alpha_{21} &= \exp\{[(A_1 + \mu_S^0) - A_2]/kT\} = K_{12}^* \\ \alpha_{31}/\alpha_{13}^* &= \exp\{[A_3 - (A_1 + \mu_P^0)]/kT\} = K_{31}^* \end{aligned} \tag{4.37}$$

and

$$\alpha_{12}^*\alpha_{23}\alpha_{31}/\alpha_{21}\alpha_{32}\alpha_{13}^* = e^{(\mu_S{}^0 - \mu_P{}^0)/kT} = K_{12}^* K_{23} K_{31}^* \tag{4.38}$$

$$\equiv K = c_P/c_S^e, \tag{4.39}$$

in agreement with Eq. 3.34. The three members of Eq. 4.38 are not equal to unity as in Eq. 4.34 (or, at equilibrium, as in Eq. 4.36), but they are independent of c_S and c_P. Figure 4.4c shows a possible infinite (and invariant) set of standard free energy levels constructed from Fig. 4.4a (dotted lines in Fig. 4.4c) on the assumption that both $c_S < 1$ M and $c_P < 1$ M in Fig. 4.4a (whereas, in Fig. 4.4c, $c_S = c_P = 1$ M). The free energy drop in one cycle is $\mu_S^0 - \mu_P^0$; this is not the actual free energy drop, $\mu_S - \mu_P$.

(b) Using this same model, we now consider another example in which we discuss entropy production and compare basic and gross free energy levels (this requires some numerical calculations—to obtain the p_i).

Figure 4.5a shows the hypothetical set of basic free energy levels we use, with nonhorizontal lines indicating possible transitions. Note that, in this example, $\Delta A'_{ij}$ is not positive in every counterclockwise step. In a particular model, any choice of α's must satisfy Eq. 4.36. Figure 4.5b gives a corresponding set of steady-state gross free energy levels (calculated as described below). As illustrated in Fig. 4.5b, the total gross free energy drop for one circuit, the sum of the $\Delta\mu_{ij}'^{\infty}$, is also $\mu_S - \mu_P$.

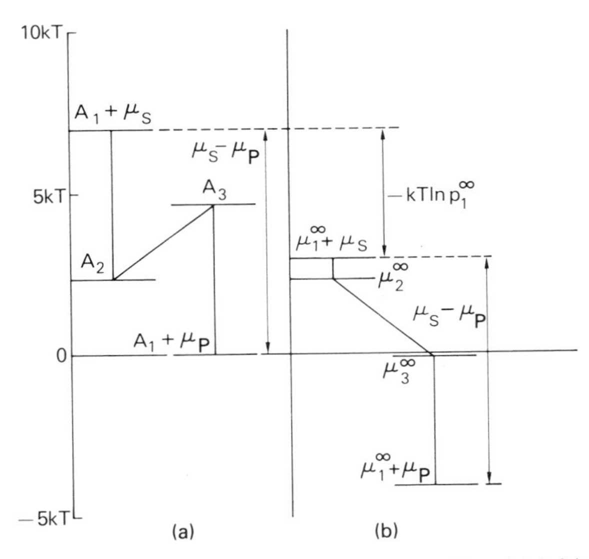

FIG. 4.5 (a) Basic free energy levels corresponding to Figs. 4.3 and 4.6. (b) Steady-state gross free energy levels for same system.

The above property of the gross free energies obtains even in a transient (i.e., at arbitrary t), as shown in Eq. 4.18. It should be pointed out, however, that in transients $\Delta\mu'_{ij}$ is not necessarily positive for every counterclockwise step. For example, suppose that, at $t = 0$, $p_3 = 1$ and $p_1 = p_2 = 0$. Then for $t > 0$ but small we would clearly have $J_{23} < 0$ and hence $\Delta\mu'_{23} < 0$.

We turn now to the steady-state situation. As there is only one cycle in the diagram, we must have $J_{12}^{\infty} = J_{23}^{\infty} = J_{31}^{\infty} \equiv J^{\infty}$ at steady state. Since

$$T\frac{d_iS}{dt} = \sum J_{ij}^{\infty}\,\Delta\mu_{ij}'^{\infty} = J^{\infty}\sum \Delta\mu_{ij}'^{\infty} = J^{\infty}(\mu_S - \mu_P) > 0 \tag{4.40}$$

and $\mu_S - \mu_P > 0$, we also have $J^\infty > 0$ (i.e., the net flux is in the direction of the force). Further, since $J^\infty \, \Delta\mu_{ij}'^\infty$ ($ij = 12, 23, 31$) > 0 (Section 4.1), we deduce that $\Delta\mu_{ij}'^\infty$ ($ij = 12, 23, 31$) > 0. Thus, at *steady state*, the gross free energy level must decrease (Fig. 4.5b) and the net flux must be positive (and equal) for each step in the direction of the force (S → P). Further, because J^∞ is constant around the cycle, the relative contribution of each transition to the overall entropy production is proportional to $\Delta\mu_{ij}'^\infty$. The results in this paragraph obviously apply to *any single-cycle model.*

There is no fundamental complication when a single cycle contains more than one force (e.g., $\mu_S - \mu_P$ above plus the force $\mu_A - \mu_B$ from the concentration gradient of a ligand, where A and B refer to the two sides of a membrane). The *net* force determines the direction of positive flux (Chapter 3).

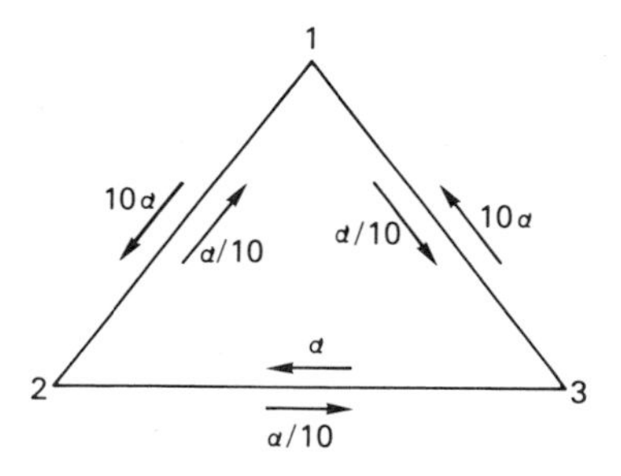

FIG. 4.6 Rate constants used in calculations for Fig. 4.5.

As a specific example, let us use the particular set of rate constants in Fig. 4.6. These have been chosen to be consistent with Fig. 4.5a, which therefore shows the basic free energy levels for this case. We have $K_{12} = 100$, $K_{23} = 0.1$, $K_{31} = 100$, and $\mu_S - \mu_P = kT \ln 1000$. Given the rate constants, we can easily calculate the steady-state probabilities (Chapters 1 and 2): $p_1^\infty = 0.01855$, $p_2^\infty = 0.972$, and $p_3^\infty = 0.00901$. Most of the enzyme accumulates in state 2 (= ES) because of the relatively small rate constants for transitions out of state 2. From Eq. 4.26, the mean flux in each step of the cycle is $J^\infty/N = 0.0882\alpha$. The gross free energy levels (Fig. 4.5b) are obtained from the basic levels (Fig. 4.5a) by subtraction of $-kT \ln p_i^\infty$ in each case. The gross level drops in each counterclockwise step, as required. The rate of free energy dissipation or entropy production, $J^\infty(\mu_S - \mu_P)$, is $0.609 N\alpha kT$.

(c) For the model in Fig. 4.3 (or any single-cycle model), $\Delta\mu_{ij}'^\infty$ and J_{ij}^∞ are both positive in every step in the direction of the force, at steady state. This is true irrespective of the arrangement of the basic free energy levels (of course there is the restraint $\sum \Delta A_{ij}' = \mu_S - \mu_P > 0$, around the cycle). Thus it would seem that, say, even an uphill sequence of basic free energy levels

($\Delta A'_{12}$ and $\Delta A'_{23}$ negative; $\Delta A'_{31}$ positive—compare Fig. 4.5a) would provide a workable model of an enzyme–substrate system. It is true that such a model would work in principle, but in practice the flux would be very small if $\mu_S - \mu_P$ is large and the uphill steps are significant (the overall "drive" is large but the transition drives $1 \to 2$ and $2 \to 3$ are negative—see Section 4.1).

As an example, suppose $\Delta A'_{12}$ and $\Delta A'_{23}$ are equal, and can be either positive or negative:

$$\Delta A'_{12} = \Delta A'_{23} \equiv \Delta A, \qquad \Delta A'_{31} = (\mu_S - \mu_P) - 2\,\Delta A. \tag{4.41}$$

Also, we define

$$z^2 \equiv e^{\Delta A/kT}, \qquad y^2 \equiv e^{(\mu_S - \mu_P)/kT} = e^{X/kT}. \tag{4.42}$$

Then $z < 1$ for uphill steps in basic free energy for $1 \to 2$ and $2 \to 3$ and $z > 1$ for downhill steps. Of course $y > 1$; usually $y \gg 1$.

For the rate constants α_{ij}, in this example, we make the simple symmetrical choices

$$\begin{aligned} \alpha_{12} &= \alpha z, & \alpha_{21} &= \alpha z^{-1}, & \alpha_{23} &= \alpha z, \\ \alpha_{32} &= \alpha z^{-1}, & \alpha_{31} &= \alpha z^{-2} y, & \alpha_{13} &= \alpha z^2 y^{-1}. \end{aligned} \tag{4.43}$$

These are designed to give

$$\alpha_{12}/\alpha_{21} = z^2, \qquad \alpha_{23}/\alpha_{32} = z^2, \qquad \alpha_{31}/\alpha_{13} = y^2/z^4$$

and

$$\Pi_+/\Pi_- = y^2 = e^{(\mu_S - \mu_P)/kT}$$

as required.

Using Eq. 1.20, we find for the steady-state flux

$$J^\infty/N\alpha = z^3(y^2 - 1)/f(z, y)$$

where (4.44)

$$f(z, y) = y^2(1 + 2z^2) + y(z + z^3 + z^5) + 2z^4 + z^6.$$

With y (i.e., $\mu_S - \mu_P$) constant, it is easy to show that $J^\infty(z)$ has a maximum when $z = y^{1/3}$. This corresponds to *equal spacing* of the basic free energy levels, $\Delta A'_{31} = \Delta A$ (with $\Delta A > 0$), as might have been expected. This is illustrated in Fig. 4.7. At this z,

$$J^\infty_{\max}/N\alpha = (y^2 - 1)/3(y^{1/3} + y + y^{5/3}). \tag{4.45}$$

For $y \gg 1$, $J^\infty_{\max}/N\alpha \to y^{1/3}/3$.

If $y \gg 1$ and $z^2 \ll 1$ (large uphill steps for $1 \to 2$ and $2 \to 3$), $J^\infty/N\alpha \to z^3$ (very small). On the other hand, for large downhill steps for $1 \to 2$ and $2 \to 3$ ($y, z \gg 1$), $J^\infty/N\alpha \to y^2/z^2(y + z)$ (also very small).

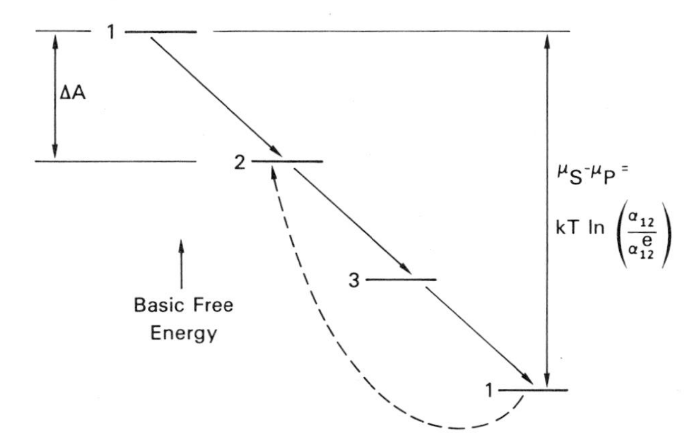

FIG. 4.7 System in Fig. 4.3 in special case of equal spacing of basic free energy levels. ΔA is defined in Eq. 4.41; α_{12}^{e} is used in Eq. 4.46. The dashed arrow corresponds to equilibrium (only the bottom three basic levels apply).

As a numerical example, let us take $y = e^{12}$ (i.e., $\mu_S - \mu_P = 24kT$, approximately as for ATP $\rightarrow$ ADP + P_i). Then J_{max}^{∞} occurs at $z = e^4$ (i.e., $\Delta A = 8kT$). Table 4.2 shows how $J^{\infty}/N\alpha$ falls off on either side of $z = e^4$. The only case included with uphill steps for $1 \rightarrow 2$ and $2 \rightarrow 3$ is $z = e^{-2}$ (i.e., $\Delta A = -4kT$). The flux is already quite small for this z.

TABLE 4.2

FLUX AND BASIC FREE ENERGY LEVEL SPACING (FOR $y = e^{12}$)

z	$J^{\infty}/N\alpha$	z	$J^{\infty}/N\alpha$
e^{-2}	0.0024	e^5	6.71
e^0	0.33	e^6	0.99
e^2	3.66	e^8	0.018
e^3	9.79	e^{10}	0.00030
e^4	18.19		

(d) Unlike the equilibrium situation, where there is a Boltzmann distribution, it is very difficult to make general statements about relative nonequilibrium populations of states. These depend, in general, on all the rate constants, the boundary conditions, and the time t. We use the present model, at steady state and at equilibrium, as an illustration.

The equilibrium case in Fig. 4.3 is achieved (we assume) by changing $\alpha_{12} = \alpha_{12}^{*} c_S$ to a much smaller value $\alpha_{12}^{e} = \alpha_{12}^{*} c_S^{e}$, i.e., $c_S \gg c_S^{e}$, without altering the other five rate constants. Thus, c_P is held constant while c_S may be varied. The equilibrium value α_{12}^e is determined by Eq. 3.31.

At equilibrium, from $\alpha_{21} p_2^e = \alpha_{12}^e p_1^e$, etc., we find

$$p_1^e \sim \alpha_{21}\alpha_{32}, \qquad p_2^e \sim \alpha_{12}^e \alpha_{32}, \qquad p_3^e \sim \alpha_{12}^e \alpha_{23}, \tag{4.46}$$

while at steady state the probabilities are given by Eqs. 1.10–1.12.

In the special case $\alpha_{12}/\alpha_{21} = \alpha_{23}/\alpha_{32} = \alpha_{31}/\alpha_{13}$, we have the (equally spaced) basic free energy level arrangement in Fig. 4.7. The dashed arrow corresponds to equilibrium (α_{12}^e). See also Figs. 4.4a and 4.4b.

As a numerical example, suppose $\alpha_{12} = \alpha_{23} = \alpha_{31} = 1$, $\alpha_{21} = \alpha_{32} = \alpha_{13} = 10^{-2}$, and $\alpha_{12}^e = 10^{-6}$ (i.e., $\alpha_{12}/\alpha_{12}^e = 10^6$). Note that Eq. 3.31 is satisfied. Then, at equilibrium (α_{12}^e), the state probabilities, in increasing order, are

$$p_2^e = 10^{-4}, \qquad p_3^e = 10^{-2}, \qquad p_1^e = 0.99, \tag{4.47}$$

with state 1 by far the most populated in this Boltzmann distribution (compare Fig. 4.7). But at steady state (α_{12}) the three probabilities are equal:

$$p_2^\infty = p_3^\infty = p_1^\infty = \tfrac{1}{3}. \tag{4.48}$$

The steady-state flux is the same along each line of the cycle:

$$J^\infty/N = \alpha_{12} p_1^\infty - \alpha_{21} p_2^\infty = 0.33. \tag{4.49}$$

As another numerical example, let us retain $\alpha_{12}/\alpha_{21} = \alpha_{23}/\alpha_{32} = \alpha_{31}/\alpha_{13}$, $\alpha_{12} = 1$, $\alpha_{21} = 10^{-2}$, and $\alpha_{12}^e = 10^{-6}$, as before, but take $\alpha_{23} = 2$, $\alpha_{32} = 2 \times 10^{-2}$, $\alpha_{31} = 4$, and $\alpha_{13} = 4 \times 10^{-2}$. That is, the transitions $2 \rightleftarrows 3$ and $3 \rightleftarrows 1$ are now somewhat faster than in the first example. The basic free energy levels (Fig. 4.7) and equilibrium probabilities (Eqs. 4.47) are unaltered, but we find, at steady state,

$$\begin{gathered} p_3^\infty = 0.15, \qquad p_2^\infty = 0.28, \qquad p_1^\infty = 0.57, \\ J^\infty/N = 0.57. \end{gathered} \tag{4.50}$$

There is some accumulation of probability at state 1 because $1 \to 2$ is a slower transition than either $2 \to 3$ or $3 \to 1$. Also, the flux is increased.

Essentially the same *steady-state* results are obtained in these examples if we neglect back reactions (to give a one-way cycle), in which case

$$p_1^\infty \sim \alpha_{12}^{-1}, \qquad p_2^\infty \sim \alpha_{23}^{-1}, \qquad p_3^\infty \sim \alpha_{31}^{-1}. \tag{4.51}$$

With three steps in the cycle and $\alpha_{12}/\alpha_{12}^e = 10^6$, the average ratio α_{ij}/α_{ji} in each step is 10^2 (because $\Pi_+/\Pi_- = \alpha_{12}/\alpha_{12}^e$). In the next example, incidentally, with *seven* states in the cycle, the corresponding numbers are 7.63×10^9 and 25.81, respectively. Both of these last two numbers are measures of the *overall* drive around the cycle.

Seven-state Myosin ATPase Cycle

In this final example (5) of a single-cycle model, we work with a complete set of recently determined (15–17) experimental rate constants for the hydrolysis of ATP by myosin S1 (which we denote by M). There are seven states in the cycle, including conformational changes and sequential release of P_i and ADP.

The practical implications for muscle contraction are discussed elsewhere (5). Also, Section 5.5 is somewhat related. We confine ourselves here to comments of a didactic nature.

The net result of the cyclic activity is the splitting of ATP in solution. No work is done and there is no free energy transduction involved.

In order to complete the set of rate constants (15–17) α_{ij}, α_{ji} (where $j = i + 1$) given in Table 4.3, we have used $10^8\ M^{-1}\ sec^{-1}$ for the three diffusion-controlled second-order binding constants α^*_{12}, α^*_{65}, and α^*_{71} and have taken for substrate and product concentrations $c_{ATP} = 3$ mM, $c_{ADP} = 0.03$ mM, and $c_{P_i} = 1$ mM.

The overall free energy changes implicit in Table 4.3 are (compare Eqs. 4.36 and 4.38)

$$\exp(\Delta\mu_{ATP}/kT) = K_{12}K_{23} \cdots K_{71} = 7.627 \times 10^9 \tag{4.52}$$

$$\exp(\Delta\mu^0_{ATP}/kT) = K^*_{12}K_{23}K_{34}K_{45}K^*_{56}K_{67}K^*_{71} = 7.627 \times 10^4 \mathrm{M}, \tag{4.53}$$

where

$$\Delta\mu_{ATP} \equiv \mu_{ATP} - \mu_{ADP} - \mu_{P_i}, \qquad \Delta\mu^0_{ATP} \equiv \mu^0_{ATP} - \mu^0_{ADP} - \mu^0_{P_i}.$$

The numerical values in Eqs. 4.52 and 4.53 correspond to

$$\begin{aligned} \Delta\mu_{ATP} &= \sum \Delta A'_{ij} = 22.755kT = 13.26 \quad \mathrm{kcal\ mol^{-1}}\ (20°\mathrm{C}) \\ \Delta\mu^0_{ATP} &= 11.242kT = 6.55 \quad \mathrm{kcal\ mol^{-1}}\ (20°\mathrm{C}). \end{aligned} \tag{4.54}$$

The remainder of Table 4.3 is largely self-explanatory. The sum of $\Delta A'_{ij}/kT$ or $\ln K_{ij}$ is 22.755 while the sum of $\ln K^*_{ij}$ (with some $\ln K_{ij}$) is 11.242, both as required. The state probabilities p_i^∞ have been calculated by computer solution of the linear algebraic equations (see Eq. 1.4). The value of p_4^∞ is relatively large because α_{45} is relatively small (the rate determining step). The p_i^∞ and $\Delta A'_{ij}/kT$ values give the $\Delta\mu'^\infty_{ij}/kT$ entries in the table, using Eq. 4.15.

Figure 4.8a presents the standard, basic, and gross free energy levels for one cycle, based on Table 4.3. There are indefinite sets of levels above and below those shown. For convenience, state 1 is set at the same free energy value in the three parts of the figure (absolute values have no significance). The overall decrease in the basic level in one cycle (the thermodynamic

TABLE 4.3

MYOSIN ATPASE CYCLE AT 20°C

State i	M + ATP 1		M · ATP 2		M* · ATP 3		M** · ATP 4	
α_{ij} (sec^{-1})		3×10^5		400		160.0		0.060
α_{ji} (sec^{-1})		2.222×10^5		2.286×10^{-6}		17.78		3.846×10^{-3}
K_{ij}		1.35		1.75×10^8		9.0		15.60
$\Delta A'_{ij}/kT = \ln K_{ij}$		0.300		18.980		2.197		2.747
K^*_{ij}		4.5×10^2 M^{-1}						
$\ln K^*_{ij}$		6.109						
p_i^{∞}	9.55×10^{-5}		1.29×10^{-4}		0.0957		0.8582	
$\Delta\mu'^{\infty}_{ij}/kT$		0.002		12.369		0.003		7.793

State i	M* · ADP · P_i 5		M* · ADP + P_i 6		M · ADP + P_i 7		M + ADP + P_i 1
α_{ij} (sec^{-1})		7.30×10^5		1.4		2.70×10^4	
α_{ji} (sec^{-1})		1×10^5		400		3×10^3	
K_{ij}		7.30		3.50×10^{-3}		9.00	
$\Delta A'_{ij}/kT = \ln K_{ij}$		1.988		−5.655		2.197	
K^*_{ij}		7.30×10^{-3} M				2.70×10^{-4} M	
$\ln K^*_{ij}$		−4.920				−8.217	
p_i^{∞}	5.53×10^{-3}		0.0403		1.25×10^{-5}		
$\Delta\mu'^{\infty}_{ij}/kT$		1.28×10^{-5}		2.423		0.165	

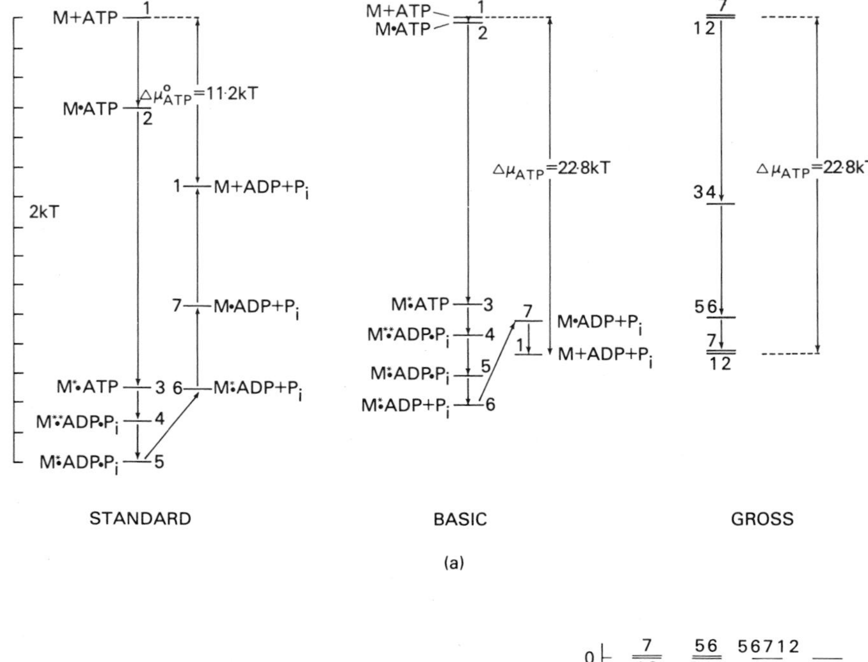

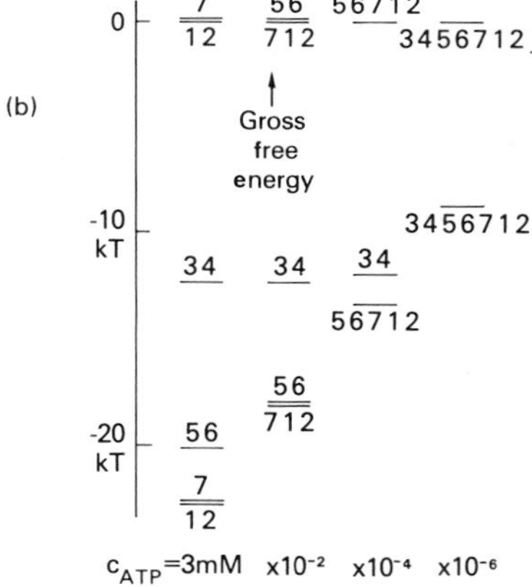

FIG. 4.8 (a) Standard, basic, and steady-state gross free energy levels for seven-state myosin-ATP cycle, based on Table 4.3. (b) Effect of varying ATP concentration on gross free energy levels.

"drive") is by no means evenly distributed: there is a very large basic free energy drop in step 2 → 3 and a significant basic free energy *increase* in step 6 → 7. The standard free energy changes are different than the basic changes for steps 1 → 2, 5 → 6, and 7 → 1 because ligand concentrations are involved.

The gross free energy level, at steady state, necessarily drops in *every* step in the dominant direction. The most striking feature of the gross free energy pattern is that three groups of states are almost in equilibrium with each other (i.e., the gross levels are nearly equal): $3 \rightleftarrows 4$; $5 \rightleftarrows 6$; and $7 \rightleftarrows 1 \rightleftarrows 2$. Thus there are effectively only three gross levels. The origin of these quasi-equilibria can be understood qualitatively from the α_{ij}, α_{ji} values in Table 4.3 (see also Appendix 1).

The entropy production arises almost entirely from transitions $2 \rightarrow 3$, $4 \rightarrow 5$, and $6 \rightarrow 7$ (see the discussion of Eq. 4.40).

An important feature to be noted in Table 4.3 is the very low p_i^∞ values for states 1, 2, and 7. Thus the equilibrium group of states 7, 1, 2 mentioned above is also a transient intermediate. Hence, for kinetic purposes, the seven-state cycle reduces, in effect, to a two-state cycle:

$$\underset{(3,4)}{\mathrm{M^* \cdot ATP, M^{**} \cdot ADP \cdot P_i}} \rightleftarrows \underset{(5,6)}{\mathrm{M^* \cdot ADP \cdot P_i\,, M^* \cdot ADP + P_i}} \rightleftarrows .$$

This two-state cycle is also shown schematically in Fig. 4.9.

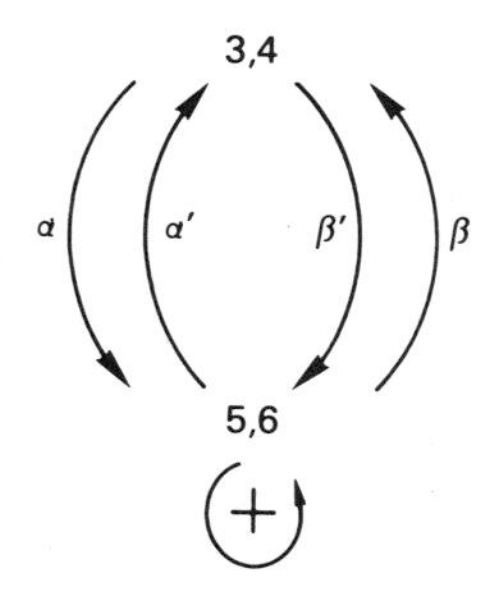

FIG. 4.9 Effective two-state cycle for system in Table 4.3 and Fig. 4.8.

The calculation of the four rate constants indicated in Fig. 4.9 provides an excellent pedagogical example of the reduction of a diagram (Appendix 1).

We first treat the three fast equilibria. Using Eq. A1.22, Eq. A1.23, and Fig. A1.3 for $3 \rightleftarrows 4$ and $5 \rightleftarrows 6$, we find from the α_{ij}, α_{ji} values in Table 4.3 the new rate constants (sec^{-1})

$$\xleftarrow{2.286 \times 10^{-7}} 3,4 \xrightarrow{0.0540} \qquad \xleftarrow{4.634 \times 10^{-4}} 5,6 \xrightarrow{1.231}. \tag{4.55}$$

Note that 0.0540 differs significantly from $\alpha_{45} = 0.0600$. Similarly, from Eqs. A1.26 and Fig. A1.4 we obtain

$$\xleftarrow{18.06} 7,1,2 \xrightarrow{219.4}. \tag{4.56}$$

Second, to handle 7, 1, 2 as a transient intermediate, we combine the above results into the reaction scheme

$$5,6 \underset{18.06}{\overset{1.231}{\rightleftarrows}} 7,1,2 \underset{2.29\times 10^{-7}}{\overset{219.4}{\rightleftarrows}} 3,4 \tag{4.57}$$

Equations A1.6 and Fig. A1.1 then give the effective rate constants

$$5,6 \underset{1.738\times 10^{-8}}{\overset{1.138}{\rightleftarrows}} 3,4 \tag{4.58}$$

with 7, 1, 2 eliminated.

Thus, from Eqs. 4.55 and 4.58, the rate constants for the two-state cycle in Fig. 4.9 are (in $\sec^{-1}$)

$$\alpha = 0.0540 \qquad \beta = 1.138 \tag{4.59}$$

$$\alpha' = 4.634 \times 10^{-4} \qquad \beta' = 1.738 \times 10^{-8}. \tag{4.60}$$

All of these rate constants are *combinations* of the elementary rate constants in Table 4.3. They lead to (Eqs. 1.7, 1.8, and 1.19)

$$p_{34}^{\infty} = 0.955, \qquad p_{56}^{\infty} = 0.0453, \qquad J^{\infty}/N = 0.0515 \quad \sec^{-1}, \tag{4.61}$$

whereas the correct values (full seven-state cycle) are

$$p_3^{\infty} + p_4^{\infty} = 0.954, \qquad p_5^{\infty} + p_6^{\infty} = 0.0458, \qquad J^{\infty}/N = 0.0514 \quad \sec^{-1}. \tag{4.62}$$

If we neglect α' and β' (one-way cycle) we again obtain essentially the results in Eqs. 4.61. We conclude that, for steady-state kinetic purposes and with the ligand concentrations chosen above, this seven-state cycle reduces with very little error to a one-way, two-state cycle.

The effect of varying ATP concentration is included in Fig. 4.8b and Table 4.9.

The rate of entropy production or free energy dissipation is (Eq. 4.40)

$$J^{\infty}\,\Delta\mu_{\mathrm{ATP}} = 0.0514 \times 13.26$$

$$= 0.682 \quad \mathrm{kcal\ mol^{-1}\ sec^{-1}}\ (20^{\circ}\mathrm{C}). \tag{4.63}$$

This example illustrates very well the fact that the standard free energy levels (Fig. 4.8a) are not relevant, except indirectly, to the actual steady-state performance of the ensemble of macromolecular units (S1 molecules in this case). The really fundamental quantities in this respect are the full set of first-order rate constants (which incorporate the actual concentration of all ligands involved). These determine both the basic free energy changes $\Delta A'_{ij}$ and the steady-state p_i^{∞}. The $\Delta A'_{ij}$ and the p_i^{∞}, in turn, determine the $\Delta\mu'^{\infty}_{ij}$. The $\Delta A'_{ij}$ per se are informative primarily in that they show how the overall drive is allotted among the various transitions. The $\Delta\mu_{ij}'^{\infty}$ indicate how close successive states are to equilibrium with each other and give, at the same time, the relative contributions of the various transitions to the rate of free energy dissipation. But the complete set of first-order rate constants contains the whole steady-state story implicitly; without the complete set, the $\Delta A_{ij}'$ and $\Delta\mu_{ij}'^{\infty}$ cannot be calculated.

The stochastic behavior of a single S1 molecule (Chapter 6) amounts to a biased one-dimensional random walk on the basic free energy levels of Fig. 4.8a (extended indefinitely, above and below). The details of such a walk, for any single molecule, do not depend at all on the condition of the ensemble (e.g., transient versus steady state). If, for example, the S1 is in state 4 at t, in the following very short time interval dt the probability that the transition $4 \rightarrow 5$ will occur is $\alpha_{45}\, dt = 0.060\, dt$, the probability of $4 \rightarrow 3$ is $\alpha_{43}\, dt = 17.78\, dt$, while the probability of no transition in dt is $1 - \alpha_{45}\, dt - \alpha_{43}\, dt$. Note that the uphill transition (Fig. 4.8a) $4 \rightarrow 3$ is more probable than the downhill transition $4 \rightarrow 5$. This emphasizes the fact that the basic free energy levels give information on relative transition probabilities only for *inverse* pairs of transitions (Eqs. 4.14). See p. 65 in this connection.

4.3 Simple Multicycle Example with Two Forces

We study here the simple model introduced in Section 2.4, Fig. 2.11, and Fig. 2.12 for the active transport of a ligand L from out (A) to in (B), against its concentration gradient, by a gradient (out $\rightarrow$ in) in Na^+. We designate $Na^+ = 1$ and $L = 2$. This model has three cycles (Fig. 2.12) and two forces and fluxes:

$$\begin{aligned} X_1 &= \Delta\mu_{\mathrm{Na}} = \mu_{1\mathrm{A}} - \mu_{1\mathrm{B}} > 0 \\ X_2 &= \Delta\mu_{\mathrm{L}} = \mu_{2\mathrm{A}} - \mu_{2\mathrm{B}} < 0, \qquad X_1 > -X_2 . \end{aligned} \tag{4.64}$$

The flux relations are

$$\begin{aligned} J_1^{\infty} &= J_a + J_b = J_{42}^{\infty} = J_{21}^{\infty} = J_{13}^{\infty} \\ J_2^{\infty} &= J_a + J_c = J_{35}^{\infty} = J_{56}^{\infty} = J_{64}^{\infty} \\ J_1^{\infty} - J_2^{\infty} &= J_b - J_c = J_{34}^{\infty} . \end{aligned} \tag{4.65}$$

The number of independent transition fluxes equals the maximum number of operational fluxes: $f = 7 - 5 = 2$. Both J_1^{∞} and J_2^{∞} are positive (out $\rightarrow$ in) in active transport. The efficiency and rate of entropy production are

$$\begin{aligned} \eta &= J_2^{\infty}(-X_2)/J_1^{\infty} X_1 < 1 \\ T\, d_i S/dt &= J_1^{\infty} X_1 - J_2^{\infty}(-X_2) > 0. \end{aligned} \tag{4.66}$$

As regards sets of basic and gross free energy levels, one can make the following general comments in advance. If there is only one force, whether there is only one or more than one cycle in the diagram, a single, infinite set of free energy levels suffices in order to follow the free energy of the system (see Sections 4.2 and 4.4). This is also sufficient if there are two or more forces but only one cycle in the diagram (because of complete coupling, there is, in effect, only one *net* force). However, if there are two forces and more than one cycle, as in the present model, a single set of free energy levels is no longer adequate to monitor the free energy. In general one needs a separate set of levels for each cycle.

Four of the seven lines in the diagram involve ligands:

$$\begin{aligned}
\alpha_{13}/\alpha_{31} &= \exp\{[(A_1 + \mu_{1A}) - A_3]/kT\} = K_{13} \\
\alpha_{35}/\alpha_{53} &= \exp\{[(A_3 + \mu_{2A}) - A_5]/kT\} = K_{35} \\
\alpha_{64}/\alpha_{46} &= \exp\{[A_6 - (A_4 + \mu_{2B})]/kT\} = K_{64} \\
\alpha_{42}/\alpha_{24} &= \exp\{[A_4 - (A_2 + \mu_{1B})]/kT\} = K_{42}\,.
\end{aligned} \tag{4.67}$$

The corresponding equations for the three isomeric transitions will be omitted. We then have, for the three cycles,

$$\Pi_{a+}/\Pi_{a-} = e^{(X_1+X_2)/kT} = K_{35}K_{56} \cdots K_{13} > 1 \tag{4.68a}$$

$$\Pi_{b+}/\Pi_{b-} = e^{X_1/kT} = K_{34} \cdots K_{13} > 1 \tag{4.68b}$$

$$\Pi_{c+}/\Pi_{c-} = e^{X_2/kT} = K_{35} \cdots K_{43} < 1. \tag{4.68c}$$

The rate constants of the model must be chosen to be consistent with the operational quantities X_1 and X_2.

As an explicit example, we now turn to the very specialized case already introduced in Fig. 2.12c and Section 2.4. In this example *all* of the counterclockwise rate constants in cycle a have the value K relative to their inverse rate constants, all of which are assigned the reference value unity. Thus all equilibrium constants in Eq. 4.68a are equal to K. The rate constants α_{34} and α_{43} (relative to the same reference value) are denoted A and B, respectively, so that $K_{34} = A/B$. Then, from Eqs. 4.68,

$$\begin{aligned}
a{:}\quad & K^6 = \exp[(X_1 + X_2)/kT] > 1, \qquad K > 1 \\
b{:}\quad & K^3(A/B) = \exp(X_1/kT) > K^6, \qquad A/B > K^3 \\
c{:}\quad & K^3(B/A) = \exp(X_2/kT) < 1.
\end{aligned} \tag{4.69}$$

The basic free energy levels for the three cycles, determined by these rate constants, are illustrated in Fig. 4.10 (an infinite set in each case). Downward steps correspond to the counterclockwise direction in cycles a and b.

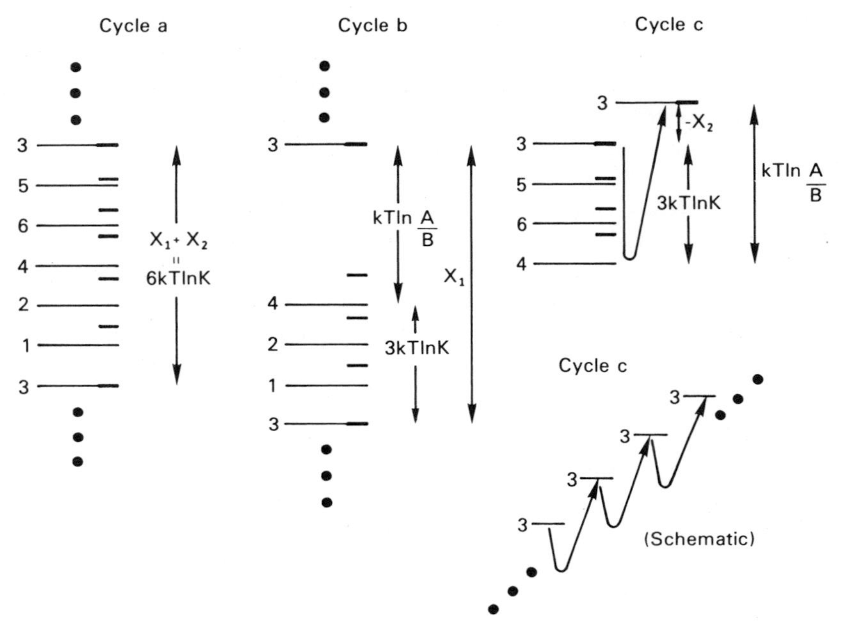

FIG. 4.10 Basic free energy levels, for Fig. 2.12c and cycles a, b, c, in the special case $A/B = K^4$. Component 1 = Na^+, component 2 = L. See text. The steady-state gross free energy levels (heavy short bars) are for the case A = 1, $B = \frac{1}{16}$, $K = 2$ (see Table 4.4). Gross and basic levels are given the same values for state 3, arbitrarily.

The single-headed arrows show the counterclockwise direction in cycle c. Some explicit choice has to be made for $A/B > K^3$; in this figure we have used $A/B = K^4$.

The basic levels are equally spaced in cycle a because all K's in Eq. 4.68a are equal. Each counterclockwise circuit around this cycle transports one L and one Na^+ from outside to inside. If this is the only significant cycle (i.e., if $A \ll 1$, $B \ll 1$), $J_1^{\mathcal{X}} = J_2^{\mathcal{X}}$ ("complete coupling") and $\eta = -X_2/X_1 = \eta_{\max} = \frac{1}{7} = 0.143$. Cycle b (34213) transports Na^+ without L. Hence, to the extent that this cycle is used, it reduces the efficiency below the above (maximum) value. The basic free energy drop per cycle b is X_1. Cycle c transports L without Na^+, but in the "wrong" direction (i.e., down the L gradient, in → out; J_c is negative). Hence, this cycle also reduces η. One counterclockwise circuit (35643) around cycle c transports one L from outside to inside and *increases* the free energy by $-X_2$ (Fig. 4.10).

It might be noted that, for the above basic free energy considerations, we need only the ratios $K/1$ and A/B and not the separate rate constants. Thus, for example, Fig. 4.10 would be unaffected if some of the six pairs of rate constants in cycle a were $10K/10$, $0.1K/0.1$, etc.

This hypothetical example illustrates the fact that free energy transduction generally has to be considered a "package" property of the *complete* mechanism or diagram: the free energy is *not* transferred in a single critical step. In one traversal of cycle a (1356421), the free energy X_1 is expended (by Na^+) and an amount $-X_2$ is recovered (by L). Where does the free energy transfer take place? The total basic free energy drop $X_1 + X_2$ in cycle a (Fig. 4.10) is broken up into six equal transition drops (which, incidentally, is kinetically optimal—as in Eq. 4.44 and Fig. 4.7). The physical processes involved (Fig. 2.11) in the sequence 1356421 are, successively, binding of Na^+, binding of L, conformational change, release of L, release of Na^+, and conformational change. The transduction of a part of X_1 into $-X_2$ is shared, inextricably, by *all* of these processes. This conclusion does *not* depend on the equal spacing of basic levels in cycle a of Fig. 4.10—this is just a simple example.

Numerical Examples

Since Eqs. 2.21–2.23, 2.27, and 2.28 for the cycle fluxes and for the p_i^∞ are available, let us pursue this example somewhat further. We use $A/B = K^4$ and $K = 2$ in the following. Then, from Eqs. 4.69, $X_1/kT = 7 \ln K = 4.852$ (this is of the correct order of magnitude for the experimental Na^+ electrochemical gradient) and $X_2/kT = -\ln K = -0.6931$.

Table 4.4 shows state probabilities, fluxes, and the efficiency for two choices of A and B, both of which are consistent with $A/B = K^4 = 16$. The states in the table are arranged in the same order as in the diagram. On the left of the table, the A and B transitions are relatively more important than on the right. Consequently, on the left, the probabilities are less uniform, cycles b and c are used more, and the efficiency is lower. If $A = B = 0$, we have all $p_i^\infty = \frac{1}{6}$ and $\eta_{max} = \frac{1}{7} = 0.143$.

TABLE 4.4

EXAMPLE: PROBABILITIES, FLUXES, EFFICIENCY[a]

$A = 1$, $B = \frac{1}{16}$ p_i^∞				$A = \frac{1}{4}$, $B = \frac{1}{64}$ p_i^∞			
(1)	0.1728	(2)	0.1974	(1)	0.1686	(2)	0.1765
(3)	0.1237	(4)	0.2097	(3)	0.1529	(4)	0.1804
(5)	0.1360	(6)	0.1605	(5)	0.1568	(6)	0.1647
$J_a/N = 0.1179$				$J_a/N = 0.1511$			
$J_b/N = 0.1040$				$J_b/N = 0.0333$			
$J_c/N = -0.00655$				$J_c/N = -0.00210$			
$J_1^\infty/N = 0.2219$				$J_1^\infty/N = 0.1844$			
$J_2^\infty/N = 0.1114$				$J_2^\infty/N = 0.1490$			
$\eta = 0.0717$				$\eta = 0.1154$			

[a] $A/B = K^4 = 16$.

The p_i^∞ in Table 4.4, for the case $A = 1$, $B = \frac{1}{16}$, have been used to calculate the gross free energy levels (by adding $kT \ln p_i^\infty$ to A_i'). These are the short heavy bars in Fig. 4.10. For convenience, these levels have been shifted vertically so that basic and gross values are the same for state 3. Since the probabilities do not vary much from state to state, the gross levels do not differ greatly from the basic levels. The gross levels for cycle c make it quite clear that, in multicycle systems at steady state, gross free energy changes in successive transitions are not always in the same direction *around each cycle*. This is in contrast to single-cycle systems (Eq. 4.40). But $\Delta\mu_{ij}'^\infty$ and J_{ij}^∞ must have the same sign for every transition (J_{ij}^∞ will generally receive contributions from more than one cycle).

The rate of entropy production is

$$\begin{aligned} T\, d_iS/dt &= J_1^\infty X_1 + J_2^\infty X_2 = (1.076 - 0.077)NkT \\ &= J_a(X_1 + X_2) + J_b X_1 + J_c X_2 \\ &= (0.490 + 0.505 + 0.005)NkT \\ &= 0.999NkT. \end{aligned} \tag{4.70}$$

The breakdown of the entropy production by transitions is given in Table 4.5.

TABLE 4.5

EXAMPLE: RATE OF ENTROPY PRODUCTION[a]

ij	$\Delta\mu_{ij}'^\infty/kT$	J_{ij}^∞/N	Product
35	0.598	0.1114	0.067
56	0.527	0.1114	0.059
64	0.426	0.1114	0.047
42	0.754	0.2219	0.167
21	0.826	0.2219	0.183
13	1.027	0.2219	0.228
34	2.244	0.1105	0.248
		Sum	0.999

[a] $K = 2$, $A = 1$, $B = \frac{1}{16}$.

Finally, we can use this example to illustrate kinetics at the cycle level, as introduced in Section 2.2. This section should be reviewed. From Eqs. 2.14 and 2.21–2.23, we have, for this example,

$$\begin{aligned} k_{a+} &= J_{a+}/N = K^6/\Sigma \\ k_{a-} &= J_{a-}/N = 1/\Sigma \\ k_{b+} &= J_{b+}/N = K^3 As/\Sigma, \end{aligned} \tag{4.71}$$

etc. These are the one-way cycle rate constants. Table 4.6 gives, on the left, numerical values for all the k's, and their net values, which are equal to J_a/N, etc. (compare Table 4.4, left side). The reciprocal of the sum of the k's is τ, the mean time between completed cycles. The cycle probabilities ($p_{a+} = \tau k_{a+}$, etc.) are on the right of Table 4.6. Of all completed cycles, 91% are of type $a+$ or $b+$. The cycle rate constants (the k's) should be compared in magnitude with the individual transition rate constants (of order unity). In fact, it is easy to show (see below) that the mean time between individual transitions in this example is $\tau_{\text{tr}} = 0.3188$, which is smaller than the mean time τ between completed cycles by a factor of 12.70. To calculate τ_{tr} we have used the following facts: the mean time $\tau_{\text{tr}}^{(i)}$ in any state i (between transitions) is the reciprocal of the sum of the *outgoing* rate constants from state i; and $p_i^\infty \sim f_i \tau_{\text{tr}}^{(i)}$, where p_i^∞ is the steady-state probability of state i and f_i is the fraction of all transitions that start from state i. The f_i can be found from the known values of the p_i^∞ and $\tau_{\text{tr}}^{(i)}$. Then $\tau_{\text{tr}} = \sum_i f_i \tau_{\text{tr}}^{(i)}$.

TABLE 4.6

EXAMPLE: CYCLE KINETICS[a]

$k_{a+} = J_{a+}/N = 0.1198$	$p_{a+} = 0.4851$
Net = 0.1179	
$k_{a-} = J_{a-}/N = 0.0019$	$p_{a-} = 0.0076$
$k_{b+} = J_{b+}/N = 0.1048$	$p_{b+} = 0.4244$
Net = 0.1040	
$k_{b-} = J_{b-}/N = 0.0008$	$p_{b-} = 0.0033$
$k_{c+} = J_{c+}/N = 0.00655$	$p_{c+} = 0.0265$
Net = −0.00655	
$k_{c-} = J_{c-}/N = 0.01310$	$p_{c-} = 0.0531$
Sum of k's = 0.2470; $\tau = 4.049$	Sum of p's = 1.0000

[a] $A = 1$, $B = \frac{1}{16}$, $K = 2$.

Directional Properties of Transitions in Multicycle Diagrams

We conclude this section with a few somewhat more general comments on Fig. 4.11. Figure 4.11 is a less specific version—the model is not definite—of Figs. 2.11 and 2.12.

We saw in the preceding section, in connection with Eq. 4.40, that it is possible to be quite explicit about directional properties of individual transitions (e.g., the sign of J_{ij}^∞) in steady-state ensembles with *single-cycle* diagrams. Multicycle diagrams, especially those with two or more forces, present a variety of possibilities. For many of these the ability to make categorical statements on steady-state directional properties is reduced to

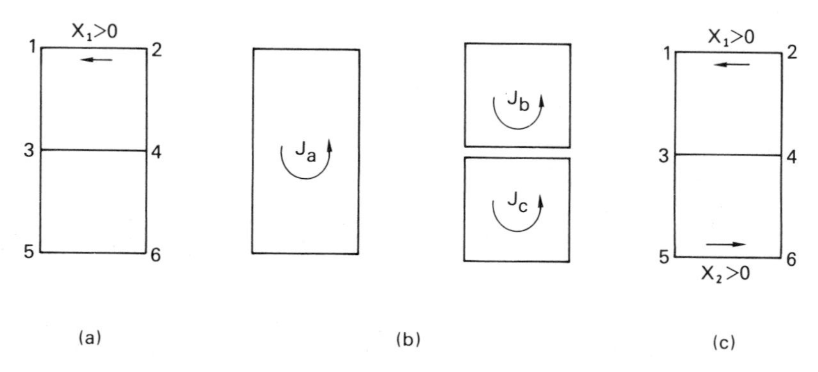

FIG. 4.11 (a) Diagram (related to Figs. 2.11 and 2.12) with single force X_1. (b) Cycles for this diagram. (c) Same diagram with two independent forces, X_1 and X_2.

some extent. We have already seen an example of this in connection with cycle c of Fig. 4.10.

With the single positive force X_1 as indicated in Fig. 4.11a,

$$\Pi_{a+}/\Pi_{a-} = e^{X_1/kT}, \qquad \Pi_{b+}/\Pi_{b-} = e^{X_1/kT}, \qquad \Pi_{c+}/\Pi_{c-} = 1. \quad (4.72)$$

Correspondingly, in Fig. 4.11a, in view of Eqs. 3.26,

$$J_a > 0, \qquad J_b > 0, \qquad J_c = 0. \quad (4.73)$$

It follows then from Eqs. 4.65 that all seven of the J_{ij}^{∞} (and $\Delta\mu_{ij}'^{\infty}$) in Eqs. 4.65 must be positive. This resembles the single-cycle situation in Eq. 4.40.

For the model with two positive forces, as shown in Fig. 4.11c, Eqs. 4.68 apply and we have

$$J_a > 0, \qquad J_b > 0, \qquad J_c > 0. \quad (4.74)$$

In this case, although the first six J_{ij}^{∞} in Eqs. 4.65 are surely positive, the sign of J_{34}^{∞} (and $\Delta\mu_{34}'^{\infty}$) is uncertain (it depends on the particular set of α's). Note that if $X_1 > X_2$ in Fig. 4.11c, we do not necessarily have $J_{34}^{\infty} > 0$, because the rate constants for the pairs of transitions 42, 21, 13 could be relatively small, leading to $J_b < J_c$. This is possible because the fluxes J_b and J_c are proportional to the differences $\Pi_{b+} - \Pi_{b-}$ and $\Pi_{c+} - \Pi_{c-}$ while the forces are related to the corresponding ratios.

4.4 Enzyme–Substrate Modified by Ligand

This example is shown in Fig. 4.12 (see also Fig. 1.2c—which is rotated 180°). An enzyme–substrate–product system is modified here by a ligand. There is one force (S → P) and no free energy transduction. An example is E = myosin (S1), S = ATP, P = ADP + P_i, L = F-actin (all in solution).

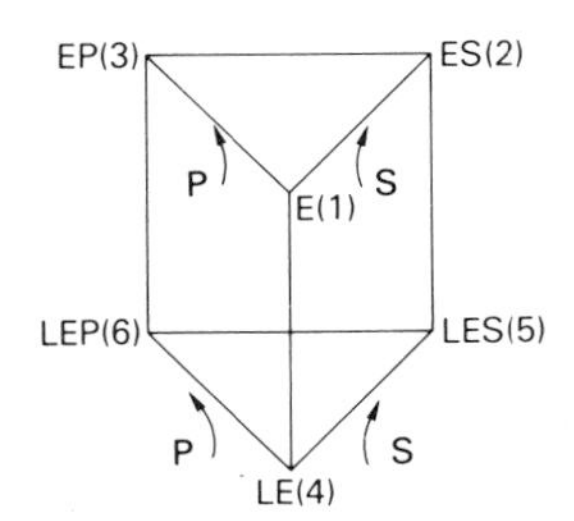

FIG. 4.12 Diagram for enzyme–substrate–product system modified by ligand L.

For this diagram, $f = 9 - 5 = 4$. We have one operational flux (S → P) and force, whereas the diagram would allow as many as four.

The diagram has 14 cycles (Fig. 4.13). Although there is only one force ($\mu_S - \mu_P > 0$), it appears at two places in the diagram. The arrow in cycles *a–h* shows both the direction of positive force and the positive direction assigned to the cycle flux. In view, then, of Eqs. 3.26 as applied to each of these cycles, we have $J_a > 0, \ldots, J_h > 0$ at steady state. Cycles *i–n* include no net force and have zero net steady-state flux. In the actin–myosin–ATP system in solution (as well as in muscle), cycle *c* is believed to dominate.

Let us begin by considering the rate of entropy production under transient conditions. This argument can obviously be extended to an arbitrary,

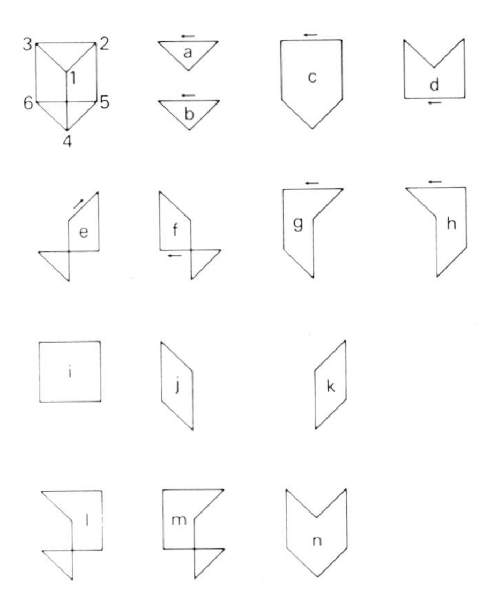

FIG. 4.13 Cycles belonging to diagram in Fig. 4.12.

more general, diagram. In Eq. 4.28, we first separate out (from $\Delta\mu'_{ij}$) the terms in μ_S, μ_P, and μ_L:

$$T\frac{d_iS}{dt} = \sum_{ij} J_{ij}\,\Delta\mu_{ij} + \mu_S(J_{12} + J_{45}) - \mu_P(J_{31} + J_{64}) + \mu_L(J_{14} + J_{25} + J_{36}), \tag{4.75}$$

where $\Delta\mu_{ij} \equiv \mu_i - \mu_j$. The ij sum is over the nine lines in the diagram. The kinetic equations of the system (Eqs. 1.3 and 1.16) can then be written

$$N\frac{dp_i}{dt} = -\sum_{k_i} J_{ik_i} \qquad (i = 1, 2, \ldots, 6) \tag{4.76}$$

where the sum here is over those (three) states k_i that are connected to state i by lines in the diagram. Using Eq. 4.76, the ij sum in Eq. 4.75 becomes

$$\sum_{ij} (J_{ij}\mu_i + J_{ji}\mu_j) = \sum_{i=1}^{6} \mu_i\left(\sum_{k_i} J_{ik_i}\right) = -N\sum_i \mu_i\frac{dp_i}{dt}. \tag{4.77}$$

The second expression results on collecting those terms that belong to each state i. Thus

$$T\frac{d_iS}{dt} = -N\sum_i \mu_i\frac{dp_i}{dt} + \mu_S(J_{12} + J_{45}) - \mu_P(J_{31} + J_{64}) + \mu_L(J_{14} + J_{25} + J_{36}). \tag{4.78}$$

Equation 4.28 expresses the rate of entropy production as a sum over the transitions of the system while Eq. 4.78 relates the same quantity to the reaction participants S, P, L, $i = 1, \ldots, 6$ [see, for example, Eq. 7.105 of Hill (18)]. At steady state, Eqs. 3.22 and 3.25 provide two other modes of expression for the rate of entropy production (but, in this example, all three expressions give the same result at steady state—see below).

We turn now to the special case of a steady state. We examine first the sign of each of the nine J_{ij}^{∞} (and hence of the $\Delta\mu_{ij}'^{\infty}$) by writing these quantities in terms of $J_a, J_b, \ldots, J_h$, all of which are positive. We can ignore $J_i, \ldots, J_n$ since they are all zero. We find, on inspection of Fig. 4.13, using simplified notation ($a \equiv J_a$, etc.),

$$\begin{aligned}
J_{12}^{\infty} &= a + d + e + g, & J_{23}^{\infty} &= a + c + g + h \\
J_{13}^{\infty} &= a + d + f + h, & J_{45}^{\infty} &= b + c + f + h \\
J_{56}^{\infty} &= b + d + e + f, & J_{64}^{\infty} &= b + c + e + g \\
J_{14}^{\infty} &= -e + f - g + h, & J_{25}^{\infty} &= -c + d + e - h \\
J_{36}^{\infty} &= c - d - f + g. & &
\end{aligned} \tag{4.79}$$

Obviously, the top six fluxes (and the $\Delta\mu_{ij}'^{\infty}$) are all positive. These are transitions in cycles *a* and *b*, in the direction of the force. But the sign of the three downward fluxes (Fig. 4.13) is uncertain. Their sum, however, is zero (because the total top and bottom populations in the diagram are constant). Although $\Delta\mu_{ij}'^{\infty}$ is positive for each step in cycles *a* and *b*, this is *not* possible for *all* of the other force-containing cycles (*c–h*). This follows from the last three of Eqs. 4.79. But of course the *sum* of $\Delta\mu_{ij}'^{\infty}$ around any of the cycles *a–h* is $\mu_S - \mu_P$. This sum is zero around cycles *i–n*.

We also note that

$$\begin{aligned} J_{12}^{\infty} + J_{45}^{\infty} \quad (\text{bind S}) &= J_{23}^{\infty} + J_{56}^{\infty} \quad (\text{S} \rightarrow \text{P}) \\ &= J_{31}^{\infty} + J_{64}^{\infty} \quad (\text{release P}) \\ &= J_a + J_b + \cdots + J_h \equiv J^{\infty}, \end{aligned} \tag{4.80}$$

where J^{∞} is the total flux in *all* cycles. Thus, at steady state, Eq. 4.78 becomes

$$T\, d_i S/dt = J^{\infty}(\mu_S - \mu_P), \tag{4.81}$$

a particularly simple result. This is a special case of both Eqs. 3.22 and 3.25.

Numerical Examples Based on Fig. 4.12

Here we supplement the preceding discussion with three numerical examples. Fig. 4.14a shows the basic free energy levels chosen for the six states. The level $A_4 + \mu_P$ is arbitrarily chosen as zero. The vertical free energy unit used (between pairs of states) is $kT \ln 300 = 5.704kT$. The thermodynamic force here, $\mu_S - \mu_P$ (double-headed arrows in Fig. 4.14), is equal to four of these units, or $22.815kT$ (roughly the free energy of ATP hydrolysis at actual concentrations). An extra level has been included at top and bottom of the figure so that all nine transitions in the diagram (Fig. 4.12) can be indicated by vertical or slanting lines. Because there is only one force, a single infinite set of free energy levels suffices (unlike Fig. 4.10).

The rate constants are shown in Fig. 4.15. This is a quite arbitrary set except, of course, inverse rate constants for each line in the diagram must be consistent with the basic free energy level difference already assigned in Fig. 4.14a (Eq. 4.14). For example,

$$\frac{\alpha_{56}}{\alpha_{65}} = \frac{\alpha}{\alpha/9 \times 10^4} = \exp\left(\frac{A_5 - A_6}{kT}\right) = (300)^2. \tag{4.82}$$

Also, the rate constants have been selected to make cycle *c* the dominant cycle (as in the myosin case).

With the rate constants available from Fig. 4.15, the steady-state probabilities of the six states may be calculated by computer from linear equations

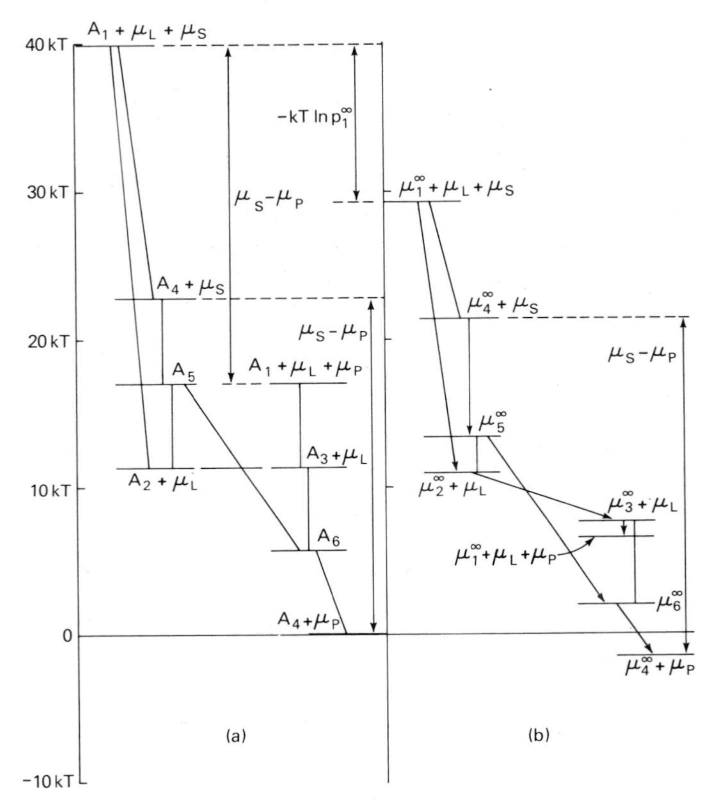

FIG. 4.14 (a) Basic free energy levels in numerical example based on Figs. 4.13 and 4.15. (b) Steady-state gross free energy levels in same example.

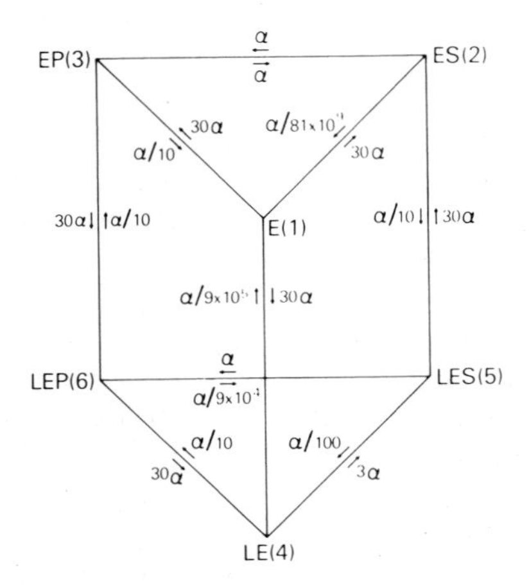

FIG. 4.15 Rate constants used in numerical example.

like 1.4 (together with $\sum_i p_i = 1$). These probabilities are given in Table 4.7. The transition fluxes $J_{ij}^{\infty}/N\alpha$ then follow from Eq. 4.26 and the cycle fluxes $J_\kappa/N\alpha$ from Eqs. 4.79. The latter calculation is possible because so many of the J_κ are zero or negligible. These fluxes are also included in Table 4.7. The important transition fluxes are those belonging to cycle c, which is seen to have the largest cycle flux by far. Cycles d–g have negligible fluxes and hence are omitted from the table. The total flux J^{∞} (Eq. 4.80) is $0.6984N\alpha$ and hence the total rate of entropy production (Eq. 4.81) is $J^{\infty}(\mu_S - \mu_P) = 15.93N\alpha kT$.

TABLE 4.7

EXAMPLE IN FIGS. 4.14 AND 4.15

i	p_i^{∞}	ij	$J_{ij}^{\infty}/N\alpha$	$J_{ij}^{\infty}\ \Delta\mu_{ij}'^{\infty}/N\alpha kT$	κ	$J_\kappa/N\alpha$
1	0.000025	12	0.00075	0.01	a	0.00075
2	0.6961	23	0.6736	2.31	b	0.0248
3	0.0225	31	0.0015	0.00	c	0.6721
4	0.2326	45	0.6976	5.54	h	0.00075
5	0.0248	56	0.0248	0.28	Sum	0.6984
6	0.0240	64	0.6969	2.39		
		14	0.00075	0.01		
		52	0.6729	1.59		
		36	0.6721	3.79		
			Sum	15.93		

Since the five large transition fluxes (those in cycle c) are all about equal, the separate contributions $J_{ij}^{\infty}\ \Delta\mu_{ij}'^{\infty}$ of these transitions to the total rate of entropy production (Table 4.7) are approximately proportional to the $\Delta\mu_{ij}'^{\infty}$. The latter quantities can be seen in Fig. 4.14b.

Figure 4.14b shows the gross free energy levels for this model, calculated from the basic levels (Fig. 4.14a) and the state probabilities in Table 4.7. That is, the ith basic level is lowered by an amount $-kT \ln p_i^{\infty}$ to obtain the ith gross level. The single-arrowed lines in Fig. 4.14b belong to those six transitions (in cycles a and b) for which $\Delta\mu_{ij}'^{\infty}$ is necessarily positive. Note that although $\Delta A_{31}'$ is negative, the steady-state probabilities p_1^{∞} and p_3^{∞} turn this around and make $\Delta\mu_{31}'^{\infty}$ positive (as it must be).

As a second example, suppose the concentration of substrate c_S is reduced by a factor of 300, with no other change in the system. This is a step nearer to equilibrium, for if c_S is reduced by a factor of $(300)^4$, then $\mu_S - \mu_P = 0$ and the steady-state kinetic system will be operating at equilibrium. In the $c_S/300$ case, all rate constants in Fig. 4.15 remain unchanged except α_{12} and α_{45}, which are both reduced by a factor of 300. In Fig. 4.14a, the basic free energy levels are the same except that the top two levels (states 1 and 4, because of the μ_S term) are both lowered by one unit, $kT \ln 300$. The left-hand side of Table 4.8 shows the new steady-state probabilities and gives the only three significant cycle fluxes. Most systems are now in state 4 because of the small rate constant α_{45}. The transition fluxes (omitted from the table) follow from Eqs. 4.79. The total flux J^∞ is reduced here to $0.00986N\alpha$ and the total rate of entropy production is lowered to $17.111kTJ^\infty = 0.169N\alpha kT$. These much smaller values reflect the fact that this steady state is closer to equilibrium than in the previous case (Figs. 4.14 and 4.15).

TABLE 4.8

EFFECT OF REDUCTION OF c_S

$c_S \times (1/300)$				$c_S \times (1/300)^4$	
i	p_i^∞	κ	$J_\kappa/N\alpha$	i	p_i^e
1	5.65×10^{-7}	b	0.00035	1	3.69×10^{-8}
2	0.00983	c	0.00949	2	1.11×10^{-5}
3	0.000328	h	0.00002	3	1.11×10^{-5}
4	0.9859	Sum	0.00986	4	0.9967
5	0.000350			5	3.69×10^{-8}
6	0.00361			6	0.00332

The gross free energy levels have the same general appearance as in Fig. 4.14b, but are somewhat compressed, with the top level ($\mu_1^\infty + \mu_L + \mu_S$) at $19.84kT$ and the bottom level ($\mu_4^\infty + \mu_P$) at $-0.01kT$.

In the equilibrium case referred to above (a third example), again only α_{12} and α_{45} are changed: both are reduced from the values in Fig. 4.15 by a factor of $(300)^4$. In this case, the top two levels in Fig. 4.14a are lowered by four units, $kT \ln(300)^4$, so that they now coincide with the levels $A_1 + \mu_P + \mu_L$ and $A_4 + \mu_P$ (since $\mu_S = \mu_P$). Thus there is only a single set of six basic free energy levels, namely, the bottom six levels in Fig. 4.14a, and there is a Boltzmann probability distribution among these levels, as given on the right-hand side of Table 4.8. State 4 is practically the only state occupied. The flux J^e and rate of entropy production are both zero. The compression of the gross free energy levels (mentioned above) is now complete, because all states are at the same gross level, $-kT \ln p_4^e = -0.0033kT$ (because we have taken $A_4 + \mu_P \equiv 0$).

If, instead of reducing the value of c_S as in the two preceding examples, we had changed the value of c_L or c_P, then the basic levels in Fig. 4.14a that include μ_L or μ_P, respectively, would have been altered, with corresponding changes in the α_{ij} (for transitions 14, 25, 36, or for 13, 46, respectively).

Table 4.9 (top) shows total flux $J^\infty/N\alpha$ as a function of force,

$$X/kT = (\mu_S - \mu_P)/kT = \ln(c_S/c_S^e),$$

where the force is varied by changing c_S (as before). The last column shows that $J^\infty/N\alpha$ is a linear function of $e^{X/kT} - 1$ out to $X/kT \cong 17$, but it is far from linear in X/kT itself over this range (see Section 3.4). A second similar example (myosin ATPase) is included in the bottom part of this table, where ATP concentration is varied.

TABLE 4.9

FLUX AS A FUNCTION OF FORCE[a]

$e^{X/kT}$	X/kT	$J^\infty/N\alpha$	$J^\infty/(e^{X/kT} - 1)N\alpha$
1	0	0	—
300^2	11.41	3.321×10^{-5}	3.69×10^{-10}
300^3	17.11	9.86×10^{-3}	3.65×10^{-10}
300^4	22.81	0.6984	0.86×10^{-10}
$e^{\Delta\mu_{ATP}/kT}$	$\Delta\mu_{ATP}/kT$	J^∞/N (sec^{-1})	$J^\infty/(e^{\Delta\mu_{ATP}/kT} - 1)N$ (sec^{-1})
1	0	0	—
7.627×10^3	8.939	1.430×10^{-5}	1.874×10^{-9}
7.627×10^5	13.545	1.401×10^{-3}	1.837×10^{-9}
7.627×10^7	18.150	3.80×10^{-2}	4.98×10^{-10}
7.627×10^9	22.755	5.14×10^{-2}	6.73×10^{-12}

[a] Top table is after Figs. 4.14 and 4.15; bottom table is after Table 4.3 and Fig. 4.8. The omitted entry in both parts of the table is the Onsager coefficient L (compare Eq. 3.70). Computing by R. M. Simmons.

Myosin–Actin–ATP Example

We provide here, for the reader interested in myosin, another example of the type just considered. This example is based on myosin kinetics in solution (19, 20) (i.e., E = myosin, S = ATP, P = ADP + P_i, L = actin). However, we emphasize that these calculations are meant to serve only as a prototype. The actual rate constants and diagram used are presumably already out of date (see, for example, Section 4.2 for a more current cycle in the absence of actin).

We use Lymn's rate constants with $c_S = 10^{-3}$M, $c_P = 10^{-5}$M,

$c_L = 5 \times 10^{-4}$M (to begin with), and a temperature of 20°C. Thus we obtain the α_{ij} and α_{ji} listed on the left side of Table 4.10 (also, we use $\alpha_{56} = \alpha_{65} = 0$). These rate constants then lead to the p_i^∞ and to the basic levels A_i' listed on the left side of Table 4.11. The symbol A_i' refers here to the level labels shown in Fig. 4.16a. That is, $A_{1S}' = A_1 + \mu_L + \mu_S$, $A_{4P}' = A_4 + \mu_P$, etc. A_{4P}' is arbitrarily chosen to have the value zero (as in Fig. 4.14). The value of $\mu_S - \mu_P$ turns out to be 10.734 kcal mol^{-1} (i.e., kT ln 10^8). Next, one can calculate J_{ij}^∞/N (Table 4.10), the gross levels $\mu_i'^\infty$ (Table 4.11), and $J_{ij}^\infty\ \Delta\mu_{ij}'^\infty$ (Table 4.10). Finally, the cycle fluxes (Table 4.11) J_κ/N can be deduced from the J_{ij}^∞/N. The total ATP flux J^∞/N is 26.18 sec^{-1} and the total rate of entropy production is 10.734 × 26.18 = 281.0 kcal mol^{-1} sec^{-1}. The corresponding numbers for the seven-state myosin cycle (no actin) in Section 4.2 are 0.0514 sec^{-1} and 0.682 kcal mol^{-1} sec^{-1}. Qualitatively, actin activates the ATPase activity of myosin, but otherwise these numbers are not strictly comparable. The free energy levels are shown in Fig. 4.16.

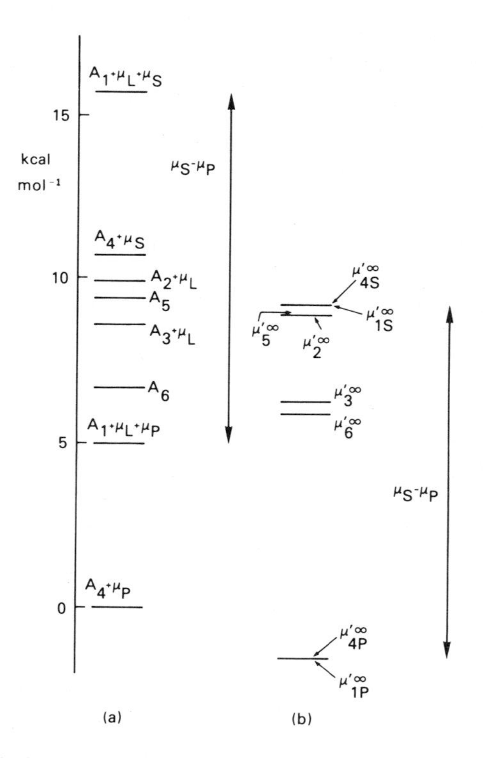

FIG. 4.16 Numerical example of same type as Fig. 4.14. This example is based on myosin–actin–ATP (Table 4.10 for rate constants). S = ATP; P = ADP + P_i; L = actin. See text. (a) Basic levels. (b) Steady-state gross levels.

TABLE 4.10

TRANSITION RATE CONSTANTS, FLUXES, ENTROPY PRODUCTION

ij	α_{ij} (sec^{-1})	α_{ji} (sec^{-1})	J_{ij}^{∞}/N (sec^{-1})	$J_{ij}^{\infty}\,\Delta\mu_{ij}'^{\infty}$ (kcal mol^{-1} sec^{-1})
12	2000	0.1	0.011	0.00
23	150	15	26.179	67.65
31	0.1	2×10^{-4}	0.002	0.02
45	1000	100	26.168	6.78
64	100	1×10^{-3}	26.177	195.11
41	5	2.5×10^{4}	0.009	0.00
52	1000	2500	26.168	0.88
36	2500	100	26.177	10.57
Total				281.01

States 2, 5, and 6 have the largest steady-state populations. Cycle c completely dominates. States 1, 2, 4 and 5 are not far from equilibrium with each other (note the values of $\mu_i'^{\infty}$ for 1S, 2, 4S, and 5). The transitions 23 and 64 contribute most to the rate of entropy production because $\Delta\mu_{23}'^{\infty}$ and $\Delta\mu_{64}'^{\infty}$ are large.

The above calculations were repeated for a thousandfold lower actin concentration, $c_L = 5 \times 10^{-7}$M. Thus, α_{14}, α_{25}, and α_{36} in Table 4.10 are all reduced by a factor of 1000. All other rate constants are unchanged. The A_i' values for states 1, 2, and 3 are all lowered by $kT \ln 10^3$. Table 4.12 includes some of the properties of this system. Values of J_{ij}^{∞}/N (omitted) are easily

TABLE 4.11

PROPERTIES OF STATES AND CYCLES

i	p_i^{∞}	A_i' (kcal mol^{-1})	$\mu_i'^{\infty}$ (kcal mol^{-1})	κ	J_κ/N (sec^{-1})
1S		15.697	9.194	c	26.166
	1.424×10^{-5}			g	0.011
1P		4.963	−1.540	h	0.002
2	0.17662	9.926	8.916	Sum	26.179
3	0.02094	8.584	6.332		
4S		10.734	9.208		
	0.97294				
4P		0.000	−1.526		
5	0.46772	9.392	8.950		
6	0.26177	6.709	5.928		

TABLE 4.12

EFFECT OF REDUCTION OF c_L ($\times 10^{-3}$)

i	p_i^∞	A_i' (kcal mol^{-1})	$\mu_i'^\infty$ (kcal mol^{-1})	κ	J_κ/N (sec^{-1})
1S		11.672	5.921	a	0.0878
	5.177×10^{-5}			c	1.1056
1P		0.938	−4.813	g	0.0060
2	0.09693	5.901	4.541	h	0.0011
3	0.88931	4.559	4.491	Sum	1.2005
4S		10.734	6.835		
	0.00124				
4P		0.000	−3.899		
5	0.00135	9.392	5.542		
6	0.01112	6.709	4.087		

obtained from the J_κ/N. The total ATP flux is reduced to 1.20 sec^{-1} and the total rate of entropy production to $10.73 \times 1.200 = 12.89$ kcal mol^{-1} sec^{-1}.

States 4, 5, and 6 are now largely unoccupied ($p_2^\infty + p_3^\infty = 0.986$). However, cycle c is still the most important cycle (though cycle a makes a significant contribution). States 2 and 3 are now almost in equilibrium with each other. The transition 64 again makes the largest contribution (8.88 kcal mol^{-1} sec^{-1}) to the rate of entropy production, because of the large value of $\Delta\mu_{64}'^\infty$.

I am indebted to Dr. R. M. Simmons for carrying out the computer calculations on this example.

Readers not interested in the subjects of Chapters 5 and 6 can proceed directly to Chapter 7 (which is independent of these two chapters).

REFERENCES

1. T. L. Hill, *Progr. Biophys. Mol. Biol.* **28**, 267 (1974).
2. T. L. Hill, *Biochemistry* **14**, 2127 (1975).
3. T. L. Hill and R. M. Simmons, *Proc. Nat. Acad. Sci. U.S.* **73**, 95 (1976).
4. T. L. Hill, and R. M. Simmons, *Proc. Nat. Acad. Sci. U.S.* **73**, 336, 2165 (1976).
5. R. M. Simmons and T. L. Hill, *Nature* (*London*) **263**, 615 (1976).
6. T. L. Hill, "Thermodynamics of Small Systems," Part I. Benjamin, New York, 1963.
7. T. L. Hill, "Thermodynamics of Small Systems," Part II. Benjamin, New York, 1964.
8. T. L. Hill, "Statistical Thermodynamics." Addison-Wesley, Reading, Massachusetts, 1960.
9. T. L. Hill, E. Eisenberg, Y. Chen, and R. J. Podolsky, *Biophys. J.* **15**, 335 (1975).
10. G. Weber, *Adv. Protein Chem.* **29**, 1 (1975).

11. A. F. Huxley, *Progr. Biophys. Mol. Biol.* **7**, 255 (1957).
12. T. L. Hill, *J. Amer. Chem. Soc.* **78**, 1577 (1956); **80**, 3241 (1958).
13. J. Keizer, *J. Theoret. Biol.* **49**, 323 (1975).
14. T. L. Hill, *Progr. Biophys. Mol. Biol.* **29**, 105 (1975).
15. C. R. Bagshaw, J. F. Eccleston, F. Eckstein, R. S. Goody, H. Gutfreund, and D. R. Trentham, *Biochem. J.* **141**, 351 (1974).
16. H. G. Mannherz, H. Schenck, and R. S. Goody, *Eur. J. Biochem.* **48**, 284 (1974).
17. D. R. Trentham, J. F. Eccleston, and C. R. Bagshaw, *Q. Rev. Biophys.* **9**, 217 (1976).
18. T. L. Hill, "Thermodynamics for Chemists and Biologists," Chapter 7. Addison-Wesley, Reading, Massachusetts, 1968.
19. R. W. Lymn, *J. Theoret. Biol.* **43**, 313 (1974).
20. R. W. Lymn, *J. Theoret. Biol.* **49**, 425 (1975).

Chapter 5 Muscle Contraction

We have already encountered examples in which the free energy of ATP splitting is used by a macromolecule to transport a ligand across a membrane against its concentration gradient. So to speak, the ATP thermodynamic force does "work" against the resisting ligand thermodynamic force. In muscle contraction (Fig. 5.1), the ligand is actin, the actin is organized into filaments, the macromolecule is a myosin cross-bridge (part of a myosin filament), the "transport" of actin takes the form of motion of the actin filaments, and the ATP thermodynamic force does mechanical work against the resisting actin filaments (which carry a load).

There is thus a close formal analogy between active transport and muscle contraction (1). The ultimate difference, as we shall see, is that the $-(J^{\infty}/N)$ $(-X)$ term for the ligand in the active transport free energy dissipation expression (see, for example, Eqs. 3.22 and 3.42) is replaced by $-(-v)(-\bar{F})$, where v is the velocity of contraction and $\bar{F}$ the average force exerted per cross-bridge on an actin filament.

A long account (2–4) of the theoretical formalism (5, 6) for the sliding-filament model of muscle contraction has been published quite recently. We give a condensed version of this formalism here. This will be followed in Section 5.3 by a discussion (7) of entropy production, directional properties, etc. Sections 5.4 and 5.5 are concerned with some recent biophysical and biochemical models.

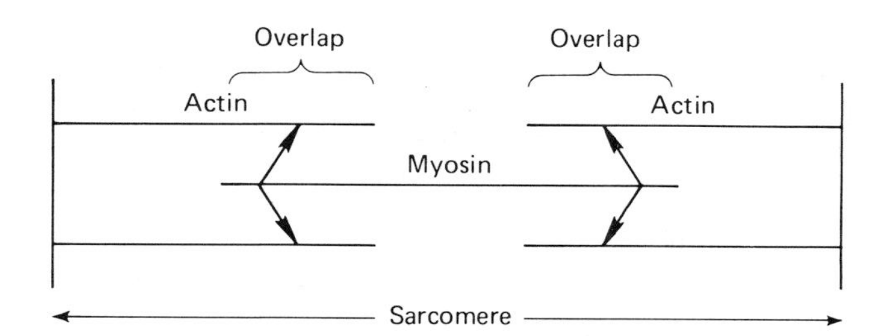

FIG. 5.1 Schematic sarcomere, showing myosin and actin filaments, overlap zones, and force exerted by cross-bridges on actin filament.

5.1 General Principles

It is assumed that the reader is acquainted with structural and other general features of the sliding-filament model of muscle contraction (8, 9).

We consider an ensemble of N independent and equivalent cross-bridges in the overlap zone (Fig. 5.1), each of which has accessible to it at most one actin attachment site at a time (3). We use the term "cross-bridge" to apply to a projection from a myosin filament whether it is attached to an actin site or not. A cross-bridge can exist in various biochemical states, as illustrated in Fig. 5.2, where M is the myosin cross-bridge, A the actin site, T = ATP, and D = ADP, P_i (treated as a single entity for simplicity). This particular

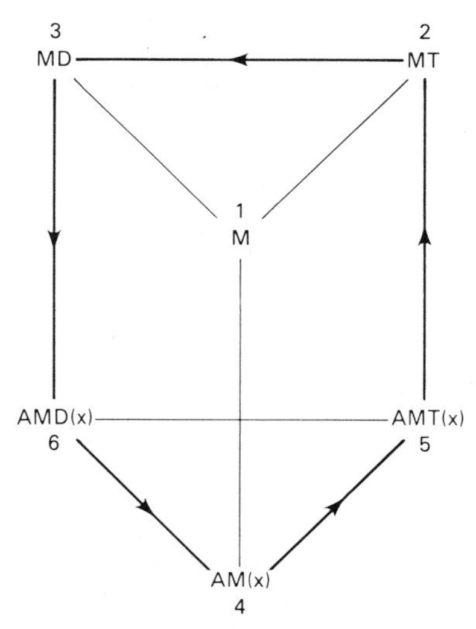

FIG. 5.2 Prototypal set of myosin biochemical states (compare Fig. 4.12). Heavy lines indicate dominant cycle. See text.

diagram is merely a prototypal example (compare Fig. 4.12). The heavy lines indicate what is currently presumed to be the dominant cycle in this diagram (10). The net flux around this cycle is in the counterclockwise direction (resulting in T → D).

Unattached and Attached States

Figure 5.3a shows, schematically, adjacent myosin and actin filaments, and one cross-bridge protruding from the myosin filament. This cross-bridge is not attached to an actin site, and no ligand (ATP, ADP, P_i) is bound to it. Its state is designated by M and its Helmholtz free energy by $A_M(T)$ or $A_1(T)$, using the numbering of states in Fig. 5.2.

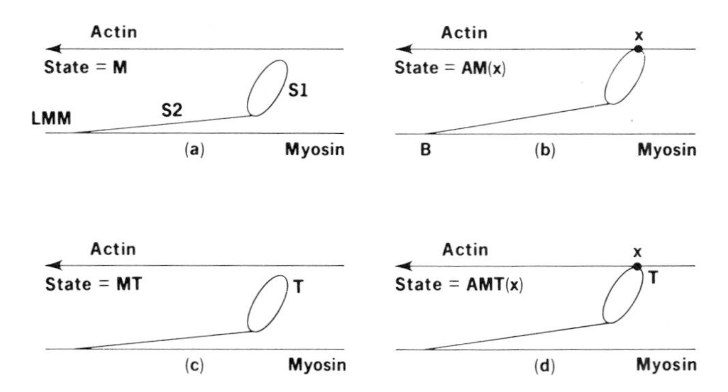

FIG. 5.3 Schematic. Four states of a myosin cross-bridge. Actin filament moves (relative to myosin) in direction of arrow in contraction. The variable x increases to the right.

In Fig. 5.3b, the cross-bridge is attached to an actin site at x, where x locates the position of the site *relative to the cross-bridge*, using some convenient origin. By convention: x increases to the right; in contraction, the actin filament moves to the left (arrow) relative to the myosin filament. The state of the cross-bridge is now designated as AM(x). The Helmholtz free energy A_{AM}, or A_4, is a function of both x and T. The variable x is essential in this problem; it is the feature that distinguishes muscle contraction from the systems treated in earlier chapters of this book.

Figures 5.3c and 5.3d illustrate, respectively, the unattached state MT(2), with free energy $A_2(T)$, and the attached state AMT(5), with free energy $A_5(x, T)$.

Each M contains two S1 molecules, or "heads." Since, for spatial reasons, presumably only one of these at a time can attach to the actin site, this feature does not complicate the preceding discussion in a formal way, but it would, of course, appear in the details of A_{AM}, etc.

Force Exerted in an Attached State

The remarks here apply to *any attached* state, which we designate below by the subscript i. Figure 5.4 shows, schematically, the cross-bridge in the attached state i at x. The force exerted by the cross-bridge, along the actin filament, is $F_i(x)$. Let this force be just balanced (for equilibrium) by an equal external force, F_i^{ext} (the analogue of the pressure exerted by the piston in fluid thermodynamics). Both of these forces are reckoned as positive in

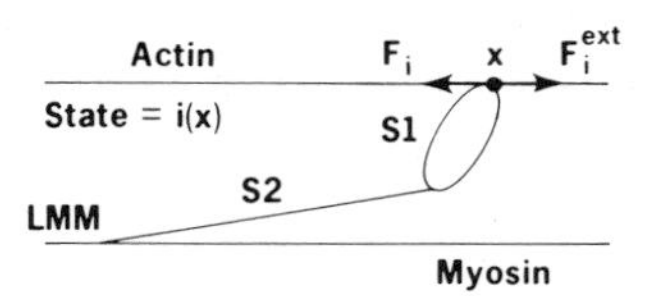

FIG. 5.4 Cross-bridge in attached state i at x. Balance of forces at actin site (see text).

the directions shown. This is a hypothetical situation with only one cross-bridge and one actin site involved in the force and thermodynamics. Now consider infinitesimal reversible changes in x and/or T. Then we have the standard relations (11, 12)

$$dA_i = -S_i\, dT + F_i\, dx, \tag{5.1}$$

$$F_i(x) = (\partial A_i/\partial x)_T \tag{5.2}$$

where $S_i(x)$ is the entropy of the cross-bridge attached at x. According to Eq. 5.2, $A_i(x)$ determines the force $F_i(x)$ exerted by the cross-bridge on the actin filament. (Of course no force is exerted on the actin filament by a cross-bridge in an unattached state such as M.) It follows from the discussion at the beginning of Section 4.1 that this force is also relevant under nonequilibrium conditions because the individual states are in "internal" equilibrium.

Assumptions of the Sliding-Filament Formalism

Having introduced the basic language, we list here, very briefly, the assumptions used (2) in the formalism to follow. Some of these assumptions have been mentioned above.

(a) At any instant a given cross-bridge has accessible to it, for attachment with significant probability, only a *single* actin site. (b) The cross-bridge behaves operationally as if it has only one head. (c) The ATP hydrolysis products ADP + P_i can be treated as a single species. (d) Cross-bridges in the overlap zone act independently of each other. (e) The kinetic and mechanical behavior of a cross-bridge is independent of the myosin–actin interfilament spacing. (f) A cross-bridge can exist in several different discrete biochemical states (providing the diagram), some attached to actin and some unattached, and transitions between these states include the binding and splitting of ATP.

Implicit in (f) are the following (see Section 4.1): the states of the diagram are in general not in equilibrium *with each other*, yet each state is an equilibrium state *internally*; transitions between states occur "instantaneously" (on the time scale of the diagram); force exerted by a cross-bridge on an actin filament is associated with *attached states* and *not with transitions*; and a nonzero force per cross-bridge (under nonequilibrium conditions) is a consequence of the effect of myofilament structural asymmetry on the rate constants of the diagram and possibly on the forces associated with attached states.

Many of the assumptions listed above can be lifted, but only at the cost of considerable complexity (3).

Basic Free Energy Changes

In the following equations we consider, for simplicity, only the main five-state cycle (heavy lines) in Fig. 5.2. In this cycle (236452), all rate constants would be expected to be functions of x except α_{23} and α_{32}. Also, A_2 and A_3 are constant while A_6, A_4, and A_5 depend on x. Thus we have

$$\begin{aligned}
\alpha_{23}/\alpha_{32} &= \exp[(A_2 - A_3)/kT] = K_{23} \\
\alpha_{36}(x)/\alpha_{63}(x) &= \exp\{[A_3 - A_6(x)]/kT\} = K_{36}(x) \\
\alpha_{64}(x)/\alpha_{46}(x) &= \exp\langle\{A_6(x) - [A_4(x) + \mu_D]\}\rangle = K_{64}(x) \\
\alpha_{45}(x)/\alpha_{54}(x) &= \exp\langle\{[A_4(x) + \mu_T] - A_5(x)\}/kT\rangle = K_{45}(x) \\
\alpha_{52}(x)/\alpha_{25}(x) &= \exp\{[A_5(x) - A_2]/kT\} = K_{52}(x).
\end{aligned} \tag{5.3}$$

The quantities that depend on x should be noted. The free energy A_i ($i = 2, 3, \ldots, 6$) refers here to a *single* independent cross-bridge (a "small" thermodynamic system) fixed in the myofilament structure, as in Eqs. 4.1 and 4.2. Thus A_i has nothing to do with the "concentration" of cross-bridges in this structure (if, say, one tries to make an analogy with free S1 or HMM molecules in solution). Note also that the "ligand" actin is not represented in these equations by μ_L terms, as would be the case for a solution system. This is because the actin site is now part of the permanent myofilament structure and is not a ligand in solution with a concentration. The role of actin is in this respect analogous to that of a solid adsorbent with sites for the binding of an adsorbate.

If we multiply Eqs. 5.3 together, around the cycle, we obtain

$$\begin{aligned}
\Pi_+(x)/\Pi_-(x) &= e^{(\mu_T - \mu_D)/kT} \\
&= K_{23}\,K_{36}(x)\,K_{64}(x)\,K_{45}(x)\,K_{52}(x),
\end{aligned} \tag{5.4}$$

where, of course, $\mu_T - \mu_D$ is a constant, independent of x. If c_T is reduced to a value c_T^e (Section 3.2) such that $\mu_T^e = \mu_D$ (equilibrium), $\alpha_{45}(x)$ (binding of T) becomes $\alpha_{45}^e(x)$ and $K_{45}(x)$ becomes $K_{45}^e(x)$. With these changes, each member of Eq. 5.4 becomes equal to unity (for any x).

Figure 5.5 shows a hypothetical set of basic free energy levels for the five-state cycle—as functions of x. These levels may be extended above and below, indefinitely. The labeling of the basic levels in such a figure is somewhat arbitrary—it depends on which set is chosen. But the differences $\Delta A'_{ij}$ are not arbitrary: these appear in Eqs. 5.3.

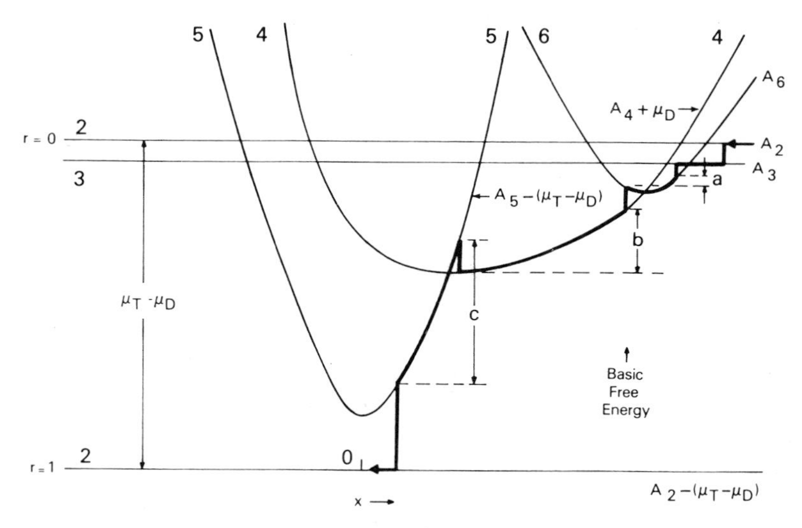

FIG. 5.5 Hypothetical set of basic free energy levels for the five-state (heavy) cycle in Fig. 5.2. The heavy path shows a possible stochastic sequence of transitions. See text for details.

Stochastics, Work, and Efficiency

Suppose we label the top five levels in Fig. 5.5 as the $r = 0$ set, the next set below [beginning with $A_2 - (\mu_T - \mu_D)$] as the $r = +1$ set, etc. Thus we have the possible sets $r = 0, \pm 1, \pm 2, \ldots$. In an *isometric contraction*, if we consider just those cross-bridges with some particular x value, any one of these cross-bridges will perform a biased random walk on the basic free energy levels at x. The transition probabilities in the walk are the $\alpha_{ij}(x)$. The general trend of the walk will be downhill, but uphill transitions will sometimes occur. If a system (cross-bridge) starts, say, in state 2 in set $r = 0$ and ends in state 2 at some $r > 0$, after many transitions, r cycles will have been completed and r molecules of ATP (in solution) will have been hydrolyzed to products. The associated free energy loss is $r(\mu_T - \mu_D)$.

Moreover, under isometric conditions (v, the velocity of contraction, equals 0), no work is performed by any of the cross-bridges (though they exert force when in attached states).

In a *steady isotonic contraction*, one can imagine a single actin site appearing on the right side of Fig. 5.5 and moving to the left side of this figure, at velocity v, past a single cross-bridge. During the "pass" of the actin site by the cross-bridge (which is in some initial state, say state 2), the cross-bridge may interact with the site via the transitions of the diagram and thus progress through some sequence of states. The transition probabilities $\alpha_{ij}(x)$ are continually changing during the pass (because x changes), as are the associated basic free energy levels (Fig. 5.5). The number of cycles completed during the pass may be $r = 0, \pm 1, \pm 2, \ldots$, with the most common result $r = 1, 2, \ldots$ (this depends on the velocity v).

The heavy path in Fig. 5.5 shows an example of a possible stochastic sequence of events in a case with $r = 1$. Transitions ("instantaneous") are represented by vertical steps, up or down, usually down. Work is done by the cross-bridge on the actin filament only when the cross-bridge is in an *attached* state. The amount of work in the example in Fig. 5.5 is $a + b + c$, these being the net basic free energy decreases of the cross-bridge while in the attached states 6, 4, and 5, respectively. The relation between work and free energy follows from (for state i)

$$\text{work in } - dx = F_i(-dx) = F_i v \, dt = -dA'_i, \tag{5.5}$$

in view of Eq. 5.2 and $v = -dx/dt$.

As far as this one pass, with $r = 1$, is concerned (Fig. 5.5), the efficiency is clearly $(a + b + c)/(\mu_T - \mu_D) =$ work done/free energy loss. The transduction of free energy (at the molecular level) is the result of the cyclic biochemical behavior of the cross-bridge, being driven by the ATP free energy drop and accomplishing work by attachment to actin. The whole cycle is involved in the transduction process, and acts as an indivisible unit. Again the analogy to active transport of a ligand by ATP, through the mediation of an enzyme, is obvious (Chapters 3 and 4).

Clearly, in order for the overall efficiency to be large, in most passes much of the total free energy drop $r(\mu_T - \mu_D)$ must be accounted for by the cross-bridge riding the basic free energy levels of attached states downward. That is, discontinuous free energy drops associated with transitions, and upward rides, must be avoided as much as possible.

The efficiency is the stochastically averaged fraction of the total free energy drop that is due to net downward rides on the free energy levels of attached states. More precisely, it is the averaged work divided by the averaged free energy drop (compare Eq. 5.19). This is not as simple a recipe for calculating the efficiency as in Eq. 5.19 below, but it does provide an intuitive and very useful way of visualizing the efficiency.

Actually, as to calculations, this stochastic approach suggests the possible use of a Monte Carlo computational treatment (2) of steady isotonic contraction as an alternative to the numerical solution of differential equations (Section 5.2). This might be the method of choice with complicated diagrams (13).

5.2 The Kinetic Formalism

Following the above introduction, we are now in a position to discuss the explicit kinetic formalism needed to make calculations on a given model. We return to the full six-state diagram in Fig. 5.2 as our prototype. In this section we designate *attached* states by subscript i ($i = 4, 5, 6$) and *any* state by subscript j or k.

Ensembles and Subensembles

We consider an ensemble (a very large number) of cross-bridges in the overlap zone. These can be divided into subensembles that contain cross-bridges with the actin site at the same x (more precisely, with the site between x and $x + dx$). Because of the lack of register between the cross-bridges on a myosin filament and the actin monomers of an actin filament (14), the distribution in the x values of the actin sites "seen" by the cross-bridges of the ensemble is uniform, or constant. Thus, all subensembles contain the same number of cross-bridges (if dx is chosen the same for each subensemble). Furthermore, this number is time invariant, irrespective of the type of experiment being considered, for if the sites are moving relative to the cross-bridges, as many cross-bridges enter the subensemble x, $x + dx$ as leave it. This assumes, of course, that both actin and myosin filaments are rigid.

The actin sites we are considering here have some repeat distance d. Since, by assumption (a) of Section 5.1, the cross-bridge can interact with only one actin site, d is presumably about 360 Å (the distance along the actin filament between actin monomers in almost equivalent orientations). In any case, a repeat distance d between eligible actin sites has to be selected as a property of any one-site model. We can regard x as locating the position of the actin site "nearest" to a given cross-bridge, with $x = 0$ the "central" position. Then there are other nonnearest actin sites at $x + d$, $x - d$, etc. The nearest site is necessarily in the range $-d/2 \leq x \leq +d/2$, with uniform probability of occurrence at any x in this interval. Thus, in averaging over x

values below, this is the appropriate interval to use, with uniform weighting in x.

Actually, an implication of the single-site assumption is that there is zero probability of cross-bridge attachment to the actin site when it is at the ends of the interval ($x = \pm d/2$). If necessary, the origin $x = 0$ should be shifted to assure this. For otherwise the cross-bridge might be attached at either $x = +d/2$ or $-d/2$, contrary to our one-site assumption.

It should perhaps be emphasized that the single-site assumption is not likely to be realistic (3), but it is used for simplicity and because all the essential features of the problem appear even in this simplified model.

Differential Equations in State Probabilities

Consider the subensemble of cross-bridges with nearest actin site between x and $x + dx$, at time t. Of these cross-bridges, let $p_j(t, x)$ be the fraction in state j at t, where $j = 1, 2, \ldots, 6$ in Fig. 5.2. Of course $\sum_j p_j = 1$ since a cross-bridge has to be in some one state at any time t.

The time evolution of the subensemble at x is governed by a set of six first-order differential equations in the p_j, only five of which are independent (because of $\sum_j p_j = 1$). For example, for state 1 (M),

$$dp_1/dt = \alpha_{21} p_2 + \alpha_{31} p_3 + \alpha_{41}(x) p_4 - [\alpha_{12} + \alpha_{13} + \alpha_{14}(x)] p_1, \quad (5.6)$$

etc. Note that α_{41} and α_{14} are x-dependent (4 is an attached state). We shall not write down the companions of Eq. 5.6 since the whole set of differential equations follows automatically from the diagram.

In Eq. 5.6, dp_1 is the change in $p_1(t, x)$ owing to transitions that occur in the infinitesimal interval dt. But, if the sites happen to be moving relative to the cross-bridges at the rate dx/dt at t (all sites necessarily move together), $p_1 + dp_1$ is the value of p_1 at $t + dt$ *and at* $x + dx$, i.e., $p_1(t + dt, x + dx)$. This provides another expression for dp_1, and Eq. 5.6 becomes

$$\left(\frac{\partial p_1}{\partial t}\right)_x + \left(\frac{\partial p_1}{\partial x}\right)_t \frac{dx}{dt} = \alpha_{21} p_2 + \alpha_{31} p_3 + \alpha_{41}(x) p_4 - [\alpha_{21} + \alpha_{13} + \alpha_{14}(x)] p_1, \quad (5.7)$$

etc. The rate dx/dt ($\equiv -v$) is the same for all j and all x, but it may be a function of t (in some transients). The left-hand side of Eq. 5.7 simplifies in various special cases, as we shall see below.

Given the rate constants and the boundary conditions (the latter are partially determined by the nature of the experiment), we can, in principle at least, solve the differential equations to obtain $p_j(t, x)$ for all j, all $t \geq 0$, and $-d/2 \leq x \leq d/2$.

In one-actin-site models, the boundary conditions would include $p_i = 0$ (attached states) at $x = \pm d/2$, $\alpha_{36} = \alpha_{14} = \alpha_{25} = 0$ (attachment rate constants) at $x = \pm d/2$, $\alpha_{63} = \alpha_{41} = \alpha_{52} = \infty$ at $x = \pm d/2$, and $p_j(d/2) = p_j(-d/2)$ (unattached states $j = 1, 2, 3$), because of the actin site periodicity (3).

The rate constants must be furnished as part of whatever molecular model is being used. Only the rate constants for transitions between unattached states (M, MT, MD) are directly related to and potentially available from solution biochemistry. The other rate constants depend not only on the molecular architecture of the filament structure but also on the particular value of x (i.e., on the position of the actin site relative to the cross-bridge).

A most important point in this connection is that the rate constants are not all independent of each other. In fact each pair α_{jk} and α_{kj} is related via equations like 5.3 to the basic free energy levels (which may have been preassigned). Further, Eq. 5.4 must be satisfied at each x. Finally, since the force functions $F_i(x)$ for attached states are also connected to the basic free energies $A_i'(x)$ by $F_i = \partial A_i'/\partial x$, care must be taken—in model building—to make the rate constants and force functions consistent with each other (4).

Force Exerted on the Actin Filament

When a cross-bridge of the subensemble at x is attached in state i to the actin site, the force exerted on the actin filament by the cross-bridge is $F_i(x)$ (Fig. 5.4), as given by Eq. 5.2. This is a consequence of the fact that state i is in internal equilibrium, as discussed in Section 4.1. $F_i(x)$ is the force exerted by the cross-bridge regardless of the nature of the experiment in progress (isotonic contraction, etc.). It is a strictly molecular property, not dependent on macroscopic external constraints such as load, etc.

The mean force (per cross-bridge) exerted on the actin filament, at t, by the cross-bridges in the subensemble at x, is then

$$F(t, x) = \sum_{i=4}^{6} p_i(t, x) F_i(x), \tag{5.8}$$

where the $p_i(t, x)$ are solutions of the differential equations above. That is, there is a contribution to this force from each *attached* state i.

The mean force (per cross-bridge) exerted on the actin filament at t by *all* cross-bridges in the ensemble is then found by averaging $F(t, x)$ over x:

$$\bar{F}(t) = \frac{1}{d} \int_{-d/2}^{+d/2} F(t, x)\, dx. \tag{5.9}$$

It is well known that a chemical reaction (T → D here) cannot generate a vector force in a homogeneous system. But the myofilament system is very asymmetric in molecular construction, so this theorem is inapplicable. Generally speaking, we should anticipate corresponding asymmetry in the functions $\alpha_{jk}(x)$ and $F_i(x)$.

Steady Isometric Contraction

We consider here the steady-state force exerted by the cross-bridges in the overlap zone on the actin filaments when the length of the muscle fiber is held constant. Then in the set of Eqs. 5.7, we put $(\partial p_j/\partial t)_x = 0$ for all j ("steady") and $dx/dt = 0$ ("isometric"). Thus, at each x, we have the linear algebraic equations

$$\begin{aligned} 0 &= \alpha_{21}p_2 + \alpha_{31}p_3 + \alpha_{41}(x)p_4 - [\alpha_{12} + \alpha_{13} + \alpha_{14}(x)]p_1 \\ 0 &= \alpha_{12}p_1 + \alpha_{32}p_3 + \alpha_{52}(x)p_5 - [\alpha_{21} + \alpha_{23} + \alpha_{25}(x)]p_2 \\ &\vdots \end{aligned} \tag{5.10}$$

which can be solved (together with $\sum_j p_j = 1$) to give each $p_{jo}(x)$ as a function of all the rate constants. (At this point we introduce the convention of designating *steady* isometric properties by a subscript letter o.) If the diagram is at all complicated, as it is in Fig. 5.2, the diagram method of Chapter 1 may be useful. Substitution of the $p_{io}(x)$ and $F_i(x)$ in Eqs. 5.8 and 5.9 then leads to $\bar{F}_o$, the mean steady isometric force generated per cross-bridge (in the overlap zone).

Equilibrium is a special case of this steady state. Equilibrium could be achieved here, in principle, by using the rate constants α_{45}^{e} and α_{12}^{e} (binding of T) appropriate to the *equilibrium* concentration c_T^e of ATP in solution (c_D being held fixed). It can be shown (2, 6) that $\bar{F}^e = 0$ at equilibrium (for any model). The nonzero steady isometric force $\bar{F}_o$ is a consequence of structural asymmetry and of the fact that the actual concentration of ATP is very much larger than its hypothetical equilibrium concentration (see below). This provides the drive for the steady-state cycling of cross-bridges within the diagram of Fig. 5.2.

Steady Isotonic Contraction

In an experiment of this kind, one fixes the load and then observes the early steady velocity of contraction, $v = -dx/dt = \text{const}$. The opposite point of view is more convenient in a theoretical calculation: the parameter v is specified in advance and the steady force generated per cross-bridge, $\bar{F}$, is then calculated from Eq. 5.9 (for the given v). This generated force is of course equal to the load (per cross-bridge in the overlap zone) that can be lifted by the muscle fiber contracting at velocity v.

In Eqs. 5.7 we put $(\partial p_j/\partial t)_x = 0$ ("steady") and $dx/dt = -v$. Since the p_j are functions of x only (and the parameter v), $(\partial p_j/\partial x)_t$ can be written dp_j/dx. Thus we have

$$-v\, dp_1/dx = \alpha_{21} p_2 + \alpha_{31} p_3 + \alpha_{41}(x) p_4 - [\alpha_{12} + \alpha_{13} + \alpha_{14}(x)] p_1, \quad (5.11)$$

etc. Steady isometric contraction is the special case $v = 0$. The solutions $p_i(x)$ of this set of equations, together with the force functions $F_i(x)$, then give $\bar{F}$ as a function of v from Eqs. 5.8 and 5.9. This is the so-called force–velocity relation (as calculated theoretically, from a model). The maximum velocity of contraction, $v_{\max}$, is the value of v that gives $\bar{F} = 0$ (no load). This property is clearly independent of the *number* of cross-bridges acting, as found experimentally (15).

An alternative but equivalent point of view in the present case is the following (compare Section 5.1 and Fig. 5.5). Instead of fixing our attention on the subensemble of cross-bridges with actin sites at x (cross-bridges move in and out of this subensemble continuously), we follow a large group of cross-bridges whose nearest actin sites all start together (as a subensemble) at $x = +d/2$ and move together at the constant velocity v toward $x = -d/2$. Thus, cross-bridges originally in the group remain in the group as time passes. The solutions of Eqs. 5.11, $p_j(x)$, starting with all $p_i = 0$ (attached states) at $x = +d/2$, can thus be thought of as describing the evolution with time (as well as x) of this group of cross-bridges as the actin sites move with constant velocity past the cross-bridges. The solutions $p_j(x)$ would agree with the mean occupations of the various states at each x if we averaged over a *very large* number of passes (all at v) in a stochastic Monte Carlo treatment of the same model (Section 5.1).

Isometric Transient

The starting point here is the steady isometric system already considered above. The cross-bridge state probabilities are $p_{jo}(x)$, the force functions are $F_i(x)$, and the mean force is $\bar{F}_o$. As in the experiments of Huxley and Simmons (16), we consider a sudden (step) change in length, from one constant value to another, amounting to y per half-sarcomere (positive y stands for stretch), so that an actin site at x relative to a cross-bridge suddenly becomes a site at $x + y$, relative to the same cross-bridge, or one at $x - y$ becomes one at x. After the step change in length but before any readjustment of the state probabilities, these probabilities are $p_j^0(x) = p_{jo}(x - y)$, where the superscript indicates the initial ($t = 0$) value of p_j at x. The final value ($t = \infty$) of p_j at x is again the steady isometric probability $p_{jo}(x)$. The final isometric steady state of the fiber differs from the original steady isometric situation (before the length change) only in the *number* of cross-bridges in the overlap zone. During the transient, $p_j^0(x) \to p_{jo}(x)$, the governing differential equations that determine the $p_j(t, x)$ are

$$(\partial p_1/\partial t)_x = \alpha_{21} p_2 + \alpha_{31} p_3 + \alpha_{41}(x) p_4 - [\alpha_{12} + \alpha_{13} + \alpha_{14}(x)] p_1, \quad (5.12)$$

etc., since $dx/dt = 0$ in Eqs. 5.7. At each x, these are linear, first-order differential equations with *constant* coefficients (unlike Eqs. 5.11). The $p_j^0(x)$ provide the boundary conditions at $t = 0$. Equation 5.9, of course, gives the mean force as a function of time.

Before the length change, and also after the transient ($t = \infty$), the mean force exerted per cross-bridge in the overlap zone is the steady isometric force $\bar{F}_o$. Immediately ($t = 0$) after the length change y, the "instantaneous" force per cross-bridge, which we denote by $\bar{F}(0; y)$, follows from Eqs. 5.8 and 5.9:

$$F(0, x + y) = \sum_{i=4}^{6} p_{io}(x)\, F_i(x + y) \quad (5.13)$$

and

$$\bar{F}(0; y) = \frac{1}{d} \int_{-d/2}^{+d/2} F(0, x + y)\, dx \quad (5.14)$$

(it is more convenient here to use $x \to x + y$ than $x - y \to x$).

Suppose (as a special case) that every $F_i(x)$ is a linear function of x, $F_i(x) = K_i(x - a_i)$, though the force constants (K_i) and locations of the zero of force $(x = a_i)$ may differ for the three attached species. Then we find from Eqs. 5.13 and 5.14 that

$$\bar{F}(0; y) = \bar{F}_o + y \sum_{i=4}^{6} K_i \left[\frac{1}{d} \int_{-\infty}^{+\infty} p_{io}(x)\, dx \right]. \quad (5.15)$$

That is, the instantaneous force $\bar{F}(0; y)$ is a linear function of y (16). However, in general, we would expect the $F_i(x)$ and hence $\bar{F}(0; y)$ to be nonlinear functions. The quantity in brackets in Eq. 5.15 is the fraction of all cross-bridges in the overlap zone that are in the attached state i (under steady isometric conditions). The experimental value of y (from the work of Huxley and Simmons) that produces $\bar{F}(0; y) = 0$ is about -40 Å.

Work and Efficiency in Steady Isotonic Contraction

In a steady isotonic contraction the force (and external load) per cross-bridge is $\bar{F}$ and the velocity of contraction is v. Since the work done (per cross-bridge in the overlap zone) in lifting the load, when the actin filaments move a distance $-dx$ relative to the myosin filaments, is $-\bar{F}\, dx$, the rate of performance of work, per cross-bridge, is simply $dW/dt = \bar{F}v$.

The free energy source for this work is the ATP in solution (around the myofilaments) at a *much* higher concentration c_T than its equilibrium concentration c_T^e. According to Eqs. 4.52 and 4.54,

$$\begin{aligned} \mu_T - \mu_D &= \mu_T - \mu_T^e \\ &= kT \ln(c_T/c_T^e) = 13.26 \quad \text{kcal mol}^{-1} \ (20°\text{C}) \end{aligned}$$

$$c_T/c_T^e = 7.63 \times 10^9, \qquad (\mu_T - \mu_D)/kT = 22.755.$$

These numbers, of course, represent an estimate. But the orders of magnitude are no doubt reliable. The system is *very* far from equilibrium.

We also need the rate at which ATP is consumed by the ensemble of systems. Consider the subensemble of cross-bridges at x. There are steady (independent of t) nonequilibrium probabilities $p_j(x)$ for the various cross-bridge states, obtained from the solution of Eqs. 5.11, and a steady net flux along each line of the diagram, Fig. 5.2. In particular, there is a net flux $J(x)$ in the two $T \rightarrow D$ transitions ($2 \rightarrow 3$, $5 \rightarrow 6$). For each (net) molecule of T converted to D, the free energy decrease in the solution is $\mu_T - \mu_D$ (since the T originates in the solution at μ_T and the D ends up in the solution at μ_D). This flux is

$$\begin{aligned} J(x) = {} & [\alpha_{23} p_2(x) - \alpha_{32} p_3(x)] \\ & + [\alpha_{56}(x) p_5(x) - \alpha_{65}(x) p_6(x)]. \end{aligned} \tag{5.16}$$

Note that this is a flux *per overlap cross-bridge*. That is, here, and *in the remainder of this chapter*, it is convenient to omit the factor N in Eq. 4.26, etc. In effect, we have already introduced this system for the force (Eq. 5.8). For the most important values of x, the term $\alpha_{23} p_2(x)$ may be much larger than the other three terms in Eq. 5.16 (note the heavy lines and arrows in Fig. 5.2). Finally, we have to average $J(x)$ over x to obtain the mean net flux per overlap cross-bridge:

$$\bar{J} = \frac{1}{d} \int_{-d/2}^{+d/2} J(x)\, dx. \tag{5.17}$$

One can show (see Eq. 5.27, below), as would be expected, that this same $\bar{J}$ is obtained for the rate of binding of T ($1 \rightarrow 2$, $4 \rightarrow 5$) or the rate of release of D ($3 \rightarrow 1$, $6 \rightarrow 4$).

Incidentally, we do not use the notation $\bar{J}^\infty$ for $\bar{J}$ because only steady isometric systems (not steady isotonic) are in steady states (at each x) of the type discussed in previous chapters. For example, the diagram method (Chapters 1 and 2) can be used only in the steady isometric case ($v = 0$).

The rate of loss of ATP free energy (per overlap cross-bridge) is then $\bar{J}(\mu_T - \mu_D)$ while the rate at which work is performed is $\bar{F}v$. Hence the rate of free energy dissipation [in ensemble plus bath plus load (2)] is

$$T\ d_iS/dt = \bar{J}(\mu_T - \mu_D) - \bar{F}v \geq 0 \tag{5.18}$$

and the efficiency is

$$\eta = \bar{F}v/\bar{J}(\mu_T - \mu_D) \leq 1. \tag{5.19}$$

Of course $\bar{F}v = 0$ either in an isometric contraction ($v = 0$) or when $v = v_{max}$ ($\bar{F} = 0$). Some intermediate v is associated with the optimal efficiency. Experimentally, the optimal efficiency (17) is of order 40–50%. Two-state, one-site models can provide efficiencies of this magnitude (4). This is an important experimental property that the more realistic models of the future (e.g., multisite, multistate) will have to contend with.

From the stochastic point of view of Section 5.1, $v/\bar{J}$ in Eq. 5.19 would be replaced by $d/\bar{r}$, where $\bar{r}$ is the mean number of ATP cycles per pass (4) at v.

A discussion of thermodynamics related to efficiency is given in Appendix III of Hill (3).

Discussion

It is worth emphasizing again that the force generated or exerted by cross-bridges on the actin filaments, and therefore the work done, is associated with biochemical *states* (specifically, attached states) and *not with transitions* or biochemical "steps." In general, each attached state will contribute to the force, though the relative magnitudes (and signs) of the contributions will depend on the probabilities of these states, the force functions, parameters such as the velocity of contraction v, and the value of x. The overall contribution of a given attached state to the force or work can be found only by averaging over x. Consideration of only one value of x will almost certainly be inadequate, even qualitatively. Averaging over x will reflect the fact that the relative importance of the different cycles and states of the diagram will ordinarily vary considerably with x. It will generally be an oversimplification or even misleading to speak of "coupling" work or load to a single state or, especially, step.

Just as the work done by the cross-bridges will usually have to be attributed in some measure to all attached states and all values of x, the free energy provided by ATP splitting cannot be localized within the diagram or at any one x. The average rate of adsorption of T, desorption of D, and $T \rightarrow D$ are all equal in magnitude and essential to the ATP turnover. But the far-from-equilibrium concentration of T in solution can be pointed to legitimately as *the* single source of free energy.

A cross-bridge with nearest actin site instantaneously at x has possible transitions and rate constants as specified by the diagram at x. But the cross-bridge has no way of "knowing" the kind of experiment being conducted. Put another way, the diagram at x is independent of the boundary (experimental) conditions that are to be used in solving the differential equations implied by the diagram. Specifically, in steady isotonic contraction, the rate constants in the diagram at any x do *not* depend on the load or force (per cross-bridge) $\bar{F}$ or on the corresponding velocity of contraction v.

$\bar{F}$ and v are necessarily properties of an *ensemble* of cross-bridges. On the other hand, the transition probabilities of a single cross-bridge at x are individual molecular properties, independent of external macroscopic (ensemble or average) conditions. A close analogy exists with an equilibrium dilute gas that is under a piston of specified weight. A particular gas molecule makes occasional collisions with the interior wall of the piston (analogous to attachment) and thereby adds its contribution to the total pressure, exerted by all the molecules, which just balances the weight of the piston. But in its detailed kinetics within the container, the molecule does not "know" what the weight is. Similarly, cross-bridges are not permanently attached to actin sites; they make independent and random "hits" (attachments) on the actin filament and their force contributions fluctuate. There is permanent continuity between half-sarcomeres in series, but this does not extend down to the cross-bridge (molecular) level.

The above two paragraphs are automatic consequences of the assumptions being used in the present formalism (2, 3, 5, 6). The assumptions, in turn, are consequences of the current common working hypothesis or "dogma" in this field. It is reasonable to expect that, in the future, these assumptions will have to be refined or generalized in various ways (3), thus requiring that some categorical statements made here be modified.

Irreversible Thermodynamics (Near Equilibrium)

The thermodynamic fluxes in Eq. 5.18 are $\bar{J}$ and $-v$ while the corresponding thermodynamic forces are $\mu_T - \mu_D$ and $\bar{F}$, respectively (18, 19). Near equilibrium, all four of these quantities are small; at equilibrium they are all equal to zero.

From a theoretical point of view (in working with a model, say), $\bar{J}$ and $\bar{F}$ are functions of the preassigned parameters $\mu_T - \mu_D$ and v. Near equilibrium, these will be linear functions:

$$\bar{F} = A(\mu_T - \mu_D) - Bv, \qquad \bar{J} = C(\mu_T - \mu_D) + Dv, \tag{5.20}$$

where A, B, C, D are positive constants (properties of the equilibrium state). The force–velocity relation, Eq. 5.20a with $\mu_T - \mu_D$ held constant, is linear near equilibrium. Under physiological conditions, this curve is far from linear (Fig. 5.6).

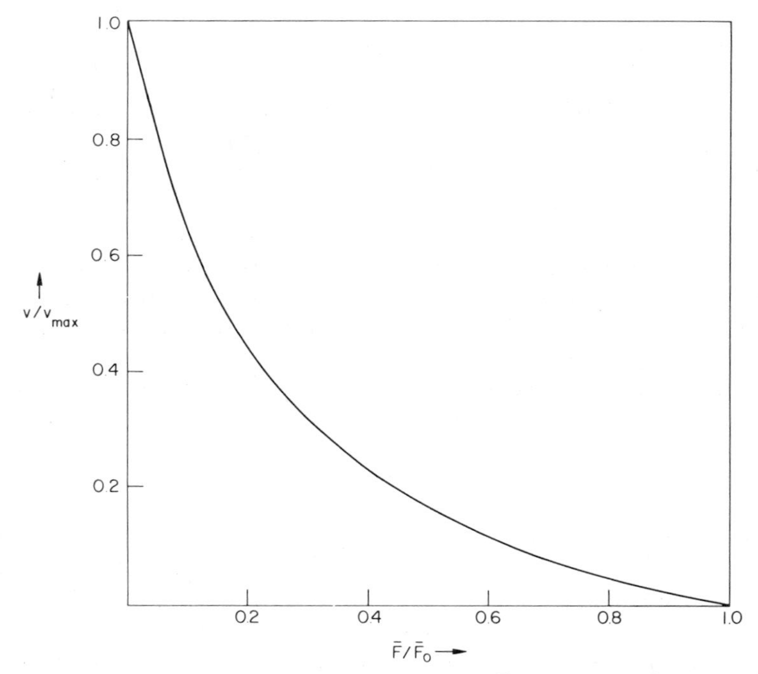

FIG. 5.6 Typical force–velocity curve: v is velocity, $\bar{F}$ the mean load per cross-bridge, v_{max} the value of v at zero load, and $\bar{F}_o$ is the value of $\bar{F}$ at $v = 0$ (isometric).

If we recast Eqs. 5.20 in the more conventional form of Eqs. 3.59, that is, fluxes as functions of forces, we find

$$\begin{aligned} \bar{J} &= [C + (DA/B)](\mu_{\text{T}} - \mu_{\text{D}}) - (D/B)\bar{F} \\ -v &= -(A/B)(\mu_{\text{T}} - \mu_{\text{D}}) + (1/B)\bar{F}. \end{aligned} \tag{5.21}$$

The reciprocal relation $L_{12} = L_{21}$ requires here that $A = D$. It has been verified in special cases (2, 4, 20, 21) and for a completely arbitrary diagram (22) that the formalism developed in this chapter does indeed lead to $A = D$. This is a self-consistency check. Generally speaking, as shown by Onsager, whenever detailed balance relations are introduced properly into a near-equilibrium theoretical system with two or more forces and fluxes, reciprocal relations will follow automatically.

Careful near-equilibrium numerical calculations on a special case have been made (20, 21), and are of considerable pedagogical value in connection with this subject.

However, it must be conceded that the physiological system itself is *very* far from equilibrium so that irreversible thermodynamics has to be relegated to the very minor role of an asymptotic (i.e., near equilibrium) test.

5.3 Entropy Production and Directional Properties

In this section (7) it will be more convenient to use subscripts i and j to represent *any* of the six states in the prototypal diagram, Fig. 5.2. We continue to omit the factor N in flux expressions (i.e., J_{ij} represents here a flux per system or per cross-bridge).

As we have seen, the free energies A_i of the unattached states ($i = 1, 2, 3$) are constants while the A_i of attached states ($i = 4, 5, 6$) are functions of x. All rate constants α_{ij} are in general functions of x except those between pairs of unattached states. For given t, $p_i(t, x)$ is the fraction of cross-bridges, with actin site at x, in state i.

We have a set of basic and a set of gross free energy levels of the sort shown in Fig. 4.14, *at each* x, where the chemical potentials μ_i are defined as (Eq. 4.2)

$$\mu_i(t, x) = A_i(x) + kT \ln p_i(t, x). \tag{5.22}$$

Note, however, as explained following Eqs. 5.3, that μ_L (actin) terms are to be omitted from Fig. 4.14 if that figure is thought of as applying here at some x (see also Fig. 5.5 where μ_L *has* been omitted).

As already mentioned following Eq. 4.28, we have here, at the outset, the fundamental directional property $J_{ij}(t, x)\, \Delta\mu'_{ij}(t, x) \geq 0$ for all transitions at any t and x. But definite directional statements about J_{ij} and $\Delta\mu'_{ij}$ separately are more difficult (as we have already seen in Sections 4.3 and 4.4, even without x dependence).

In isometric contractions, an ensemble (or subensemble) at any x has no interaction with the other x-ensembles. Therefore the conclusions in Chapter 4 apply here at each x without change. No external work is involved. Under nonisometric conditions, however, the different x-ensembles are not independent of each other and external work must be taken into account.

The interdependence of x-ensembles is implicit in the appropriate kinetic equations. These are (Eqs. 4.76 and 5.7)

$$\left(\frac{\partial p_i}{\partial t}\right)_x - v(t)\left(\frac{\partial p_i}{\partial x}\right)_t = -\sum_{k_i} J_{ik_i}(t, x) \qquad (i = 1, 2, \ldots, 6), \tag{5.23}$$

where $v(t)$ is the velocity of contraction. In a steady isotonic contraction with $v = \text{const}$, we have

$$\left(\frac{\partial p_i}{\partial t}\right)_x = 0 = v\frac{dp_i}{dx} - \sum_{k_i} J_{ik_i}(x) \qquad (i = 1, 2, \ldots, 6). \tag{5.24}$$

This equation illustrates the interaction between ensembles at different x values: Eq. 5.24 may be regarded as the steady-state kinetic equation for the components of a chemically reacting system with hypothetical *one-way* (x decreasing; $v > 0$), one-dimensional diffusion (diffusion coefficient equals v) subject to periodic boundary conditions (what "goes out" at $x = -d/2$ "comes in" at $x = +d/2$). The "diffusion" mixes the x-ensembles when $v > 0$, the more so the larger the value of v.

We see from Eq. 5.24 that, in a steady isotonic contraction, the total flux into each state i at x, owing to transitions, is *not* equal to zero; but it *is* equal to zero if we average over x:

$$0 = -\sum_{k_i} \bar{J}_{ik_i} \qquad (i = 1, 2, \ldots, 6) \tag{5.25}$$

where the term in v drops out because of the periodic boundary conditions (Section 5.2). We have used, in Eqs. 5.25, the definition (Eq. 5.17)

$$\bar{J}_{ij}(t) \equiv \frac{1}{d} \int_{-d/2}^{+d/2} J_{ij}(t, x)\, dx. \tag{5.26}$$

It is easy to show from the set of Eqs. 5.25 that

$$\bar{J}_{12} + \bar{J}_{45} = \bar{J}_{23} + \bar{J}_{56} = \bar{J}_{31} + \bar{J}_{64}, \tag{5.27}$$

just as in Eq. 4.80. But here x averaging is required and we cannot make use of the diagram method (because $v \neq 0$).

Let us turn now to the rate of entropy production *per cross-bridge*, averaged over x, and under arbitrary transient conditions:

$$T \frac{d_i S}{dt} = \frac{1}{d} \int_{-d/2}^{+d/2} \sum_{ij} J_{ij}(t, x)\, \Delta\mu'_{ij}(t, x)\, dx \geq 0, \tag{5.28}$$

where the sum is over the nine lines in the diagram (Fig. 5.2). We can separate out the terms in μ_T and μ_D (as in Eq. 4.75), use Eqs. 4.77 and 5.23, and find

$$T \frac{d_i S}{dt} = -\frac{1}{d} \int_{-d/2}^{+d/2} \sum_i \mu_i \left[\frac{\partial p_i}{\partial t} - v(t) \frac{\partial p_i}{\partial x} \right] dx + \mu_T(\bar{J}_{12} + \bar{J}_{45})$$
$$- \mu_D(\bar{J}_{31} + \bar{J}_{64}). \tag{5.29}$$

To examine the terms in $v(t)$, we replace μ_i by $A_i + kT \ln p_i$. For unattached states ($i = 1, 2, 3$), $A_i = \text{const}$. Then both integrals (A_i, $\ln p_i$) are zero (periodic boundary conditions, integration by parts). For attached states ($i = 4, 5, 6$), the $\ln p_i$ integral is also zero but the A_i integral (integration by parts) introduces the force $\bar{F}_i$, since $F_i(x) = \partial A_i(x)/\partial x$. Thus

$$T \frac{d_i S}{dt} = -\frac{1}{d} \int_{-d/2}^{+d/2} \sum_i \mu_i(t, x) \frac{\partial p_i}{\partial t}\, dx - \bar{F}(t)\, v(t)$$
$$+ \mu_T[\bar{J}_{12}(t) + \bar{J}_{45}(t)] - \mu_D[\bar{J}_{31}(t) + \bar{J}_{64}(t)] \geq 0, \tag{5.30}$$

where $\bar{F} = \bar{F}_4 + \bar{F}_5 + \bar{F}_6$. The "reaction participant" terms ($i = 1, \ldots, 6$; T, D) are essentially as in Eq. 4.78, but a new feature here is the appearance of the rate of performance of external work, $\bar{F}v$. The work term obviously originates, mathematically, from the x-ensemble "mixing" term, $-v\,\partial p_i/\partial x$, in Eq. 5.23. The interpretation of Eq. 5.30: not all of the macroscopic free energy drop in the overall reaction system [terms in μ_i, μ_T, μ_D; see Eq. 7.105 of Hill (23)] is wasted or dissipated ($T\,d_i S/dt$); some of it is converted into external work ($\bar{F}v$). Or-to turn this statement around: the macroscopic free energy drop of all participants arises not only from dissipative transitions but also from external work done by the system.

In a steady isotonic contraction, Eq. 5.30 simplifies to

$$T\,d_i S/dt = \bar{J}(\mu_T - \mu_D) - \bar{F}v \geq 0, \tag{5.31}$$

where $\bar{J}$ (ATP flux) is defined as either of the expressions in brackets in Eq. 5.30 (see Eq. 5.27). This same relation was obtained in a very different way in Eq. 5.18.

Under steady isotonic conditions, it is easy to see that we also have

$$T\,\frac{d_i S}{dt} = \frac{1}{d}\int_{-d/2}^{+d/2} \sum_{ij} J_{ij}\,\Delta A'_{ij}\,dx = \bar{J}(\mu_T - \mu_D) - \bar{F}v \geq 0 \tag{5.32}$$

because the ln p_i terms in Eq. 5.28, collectively, make no contribution to the final result. Thus, so to speak, the dissipation owing to all transitions ij, averaged over x, can be monitored either on the gross free energy levels (Eq. 5.28) or on the basic free energy levels (if $v =$ const). The latter point of view is used, in effect, in our earlier discussion of Fig. 5.5.

The contribution of the transition ij (averaged over x) to the entropy production is clearly

$$\frac{1}{d}\int_{-d/2}^{+d/2} J_{ij}\,\Delta\mu'_{ij}\,dx \geq 0. \tag{5.33}$$

But this ij contribution cannot be further separated into a work and an ATP flux term (as in Eq. 5.31 or 5.32) because work is associated with states, not transitions. Equation 5.33 is *not* always true if $\Delta\mu'_{ij}$ is replaced by $\Delta A'_{ij}$ [Hill and Simmons (7), p. 2165]. That is, the $\Delta A'_{ij}$ inequality in Eq. 5.32 is a "collective" property that does not necessarily hold for each separate transition.

Finally, let us consider what can be said about the sign of the nine $\bar{J}_{ij}$ and $\overline{\Delta\mu'_{ij}}$ in Fig. 5.2 in a steady isotonic contraction, where, as in Eq. 5.26, we define

$$\overline{\Delta\mu'_{ij}} \equiv \frac{1}{d}\int_{-d/2}^{+d/2} \Delta\mu'_{ij}\,dx. \tag{5.34}$$

We limit the discussion to systems for which $\bar{F} \geq 0$ and $v \geq 0$ (other cases are possible).

When $v = 0$, the diagram method can be applied *at each* x, as in Eqs. 4.79 and 4.80. Thus the first six $J_{ij}(x)$ in Eq. 4.79 [and the corresponding $\Delta\mu'_{ij}(x)$] are positive at every x. On integrating over x, these six $\bar{J}_{ij}$ (and $\overline{\Delta\mu'_{ij}}$) are necessarily positive. The other three $\bar{J}_{ij}$ and $\overline{\Delta\mu'_{ij}}$ are uncertain in sign; it is even possible that one or more corresponding pairs would not agree in sign. But the sum of the three $\bar{J}_{ij}$ must be zero (as it is at every x). Just as in Section 4.4, $\overline{\Delta\mu'_{ij}}$ is positive for all the steps in cycles a and b but the same cannot be said for *all* of cycles c–h.

If $v > 0$, we have Eqs. 5.25 and 5.27, on integrating over x. Each pair $(= \bar{J})$ in Eq. 5.27 is surely positive, as a consequence of Eq. 5.31. We also have

$$\bar{J}_{14} + \bar{J}_{25} + \bar{J}_{36} = 0. \tag{5.35}$$

But the sign of the nine individual $\bar{J}_{ij}$'s is not obvious. The same is true of the nine $\overline{\Delta\mu'_{ij}}$'s. Also, $\bar{J}_{ij}$ and $\overline{\Delta\mu'_{ij}}$ need not always agree in sign. It is easy to see that the *sum* of the $\overline{\Delta\mu'_{ij}}$ around any of the cycles a–h is $\mu_T - \mu_D$. This sum is zero around the other cycles (i–n).

If $v > 0$ and only a single cycle is important, say cycle c (Fig. 4.13), 236452, then one can go further (see Eq. 5.25):

$$\bar{J} = \bar{J}_{23} = \bar{J}_{36} = \bar{J}_{64} = \bar{J}_{45} = \bar{J}_{52} > 0. \tag{5.36}$$

That is, *after* x-averaging, each step has equal and positive flux. At individual values of x, these $J_{ij}(x)$ are not only not equal (Eq. 5.24) but negative values are possible. For example, consider $J_{52}(x)$ near $x = +d/2$ when $v > 0$. Because $p_2 \neq 0$ and $p_3 \neq 0$ while $p_4 = p_5 = p_6 = 0$ (attached states) at $x = +d/2$, the rate of the transition $5 \to 2$ is zero. Therefore both $J_{52}(x)$ and $\Delta\mu'_{52}(x)$ must be negative near $x = +d/2$. The larger v the greater the x interval over which these negative values will persist before becoming positive. These comments obviously resemble those made in Section 4.2 concerning a transient with t near $t = 0$.

In the single-cycle, $v > 0$ case, the sum of the $\overline{\Delta\mu'_{ij}}$ around the cycle is $\mu_T - \mu_D$. Despite Eq. 5.36, it is not certain that all of the individual $\overline{\Delta\mu'_{ij}}$'s are positive. For example, it seems possible that $\overline{\Delta\mu'_{52}}$ might be negative when v is large. Of course all of the $\overline{\Delta\mu'_{ij}}$ must be positive if $v = 0$.

As can be seen from the preceding discussion, x-averaging of J_{ij} and $\Delta\mu'_{ij}$ (at arbitrary v) does not lead to any particularly simple or fundamental thermodynamic relations involving $\bar{J}_{ij}$ and $\overline{\Delta\mu'_{ij}}$. For example, the ij entropy production in Eq. 5.33, $\overline{J_{ij}\,\Delta\mu'_{ij}}$, is not equal to $\bar{J}_{ij} \cdot \overline{\Delta\mu'_{ij}}$. Although $\bar{J}_{ij}$ is an operational quantity, $\overline{\Delta\mu'_{ij}}$ does not seem to have any fundamental significance. Correspondingly, there does not appear to be any correct way of giving an x-averaged kinetic account of this system (i.e., using x-averaged rate constants, state probabilities, etc.). Many of the fundamental parameters of a model are functions of x, and this level of detail cannot be escaped in a rigorous treatment.

Numerical illustration of free energy levels and entropy production, using an explicit model of muscle contraction at different values of v, has not been carried out as yet but is planned for the future (Simmons and Hill).

5.4 Current Status of Muscle Models

We mention only complete and self-consistent models here, which are based on the above formalism.

Work is really only just beginning on this subject. Models that are fairly realistic both biochemically and structurally are being worked on currently by A. F. Huxley and Simmons and by Eisenberg, Hill, and Chen (both unpublished as yet). Both groups are using one or two unattached states and two or three attached states, plus *multiple* actin sites (55 Å apart). These are all generalizations of the preliminary Huxley–Simmons (16) model. For example, if there are four biochemical states and four sites to be considered, this makes a total of 16 different states that must be contended with in the diagram and formalism (3). Huxley and Simmons are, of course, adjusting the parameters of their model to fit, as well as possible, the extensive experimental data they have accumulated (to be published), particularly on isometric transients.

Eisenberg and Hill (24) have given a qualitative discussion of the essential biochemical and biophysical features involved in models of the above type. Some of the theoretical principles that relate to individual rate constants of the myosin–actin–ATP system have been discussed by Hill and Eisenberg (25) and Hill (26, 27).

Steady properties of a number of self-consistent and complete *two-state* models were calculated by Hill *et al.* (4). These models were based on the earlier work of Podolsky and Nolan (28). The primary motivation was to test the formalism and check orders of magnitude. The results were reassuring. For example, it is quite apparent that models based on the formalism outlined in this chapter can give thermodynamic efficiencies of the order of 40–50%, as required.

Other much less realistic two-state models were used (20, 21) to explore *theoretical* properties over the complete (ATP concentration) range of very near equilibrium to very far from equilibrium.

The above three papers (4, 20, 21) comprise essentially an appendix to the present chapter.

5.5 Free Energy Transfer in Muscle Contraction

Because the biochemistry of the myosin–actin–ATP system is relatively well understood, we can discuss the free energy transfer in muscle contraction, in molecular terms, at least qualitatively (24).

Figure 5.7a shows the single-site, four-state cycle we consider (24). The superscripts on M refer to different conformations. The transition $4 \rightarrow 1$ is actually a composite of several steps, with transient intermediates, as indicated in Fig. 5.7b. The presumed qualitative arrangement of the basic free energy curves (2, 16, 24) is similar to Fig. 5.5 (five states), and is shown in Fig. 5.7c. The vertical arrows indicate typical stochastic transition points

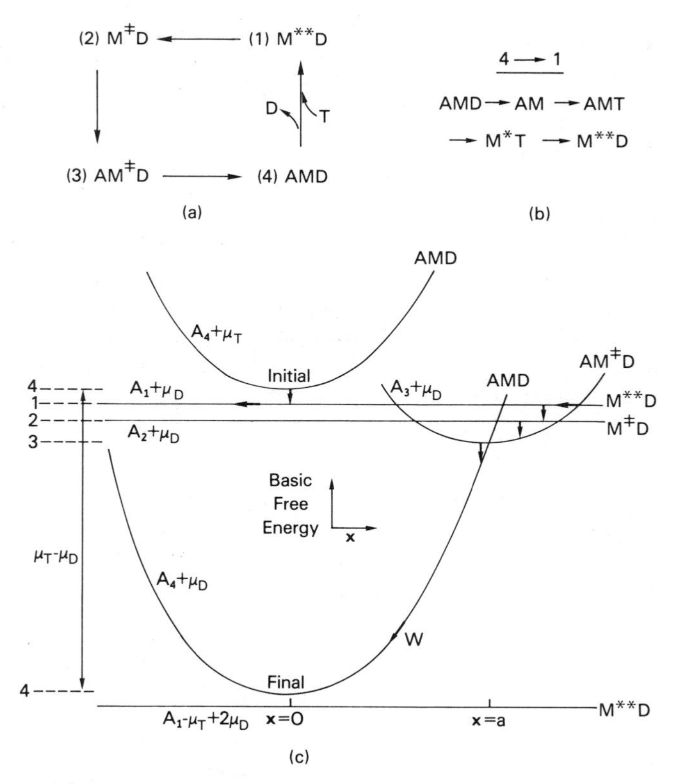

FIG. 5.7 (a) Four-state cycle used in discussion of free energy transfer in muscle. (b) Details (transient intermediates) of transition $4 \rightarrow 1$, AMD $\rightarrow$ M**D. (c) Qualitative arrangement of basic free energy curves for the four-state cycle. Short, heavy arrows show a more or less optimal stochastic sequence in one cycle, beginning at the top ("initial") in state AMD. Work *W* is accomplished at the end of the cycle, in state AMD.

(compare Fig. 5.5). If the system operated in homogeneous solution, the basic free energy levels would be essentially as shown on the left of Fig. 5.7c (dashed lines).

We consider a cycle starting in state 4 (AMD) at $x \cong 0$ and ending in the same state:

$$\begin{gathered} \mathrm{AMD}\ (x \cong 0) + \mathrm{T} \rightarrow \mathrm{AMD}\ (x \cong 0) + \mathrm{D} \\ A_4\ (x \cong 0) + \mu_\mathrm{T} \rightarrow A_4\ (x \cong 0) + \mu_\mathrm{D}. \end{gathered} \tag{5.37}$$

It is convenient to refer to the initial state as a high (basic) free energy state (relative to the final state which is lower by an amount $\mu_\mathrm{T} - \mu_\mathrm{D}$). In view of the very strong binding of T to M (29) (practically of magnitude $\mu_\mathrm{T} - \mu_\mathrm{D}$), as seen in Fig. 4.8 (M → M*T), the high free energy of the initial state of our cycle may be considered to reside primarily in the potential for strong binding of T to the M of AMD, i.e., essentially in M + T relative to M*T.

The binding of A to M is almost equally strong (30). In the process (part of 4 → 1; Fig. 5.7b)

$$\mathrm{AMD}\ (x \cong 0) + \mathrm{T} \rightarrow \mathrm{A} + \mathrm{M^*T} + \mathrm{D}, \tag{5.38}$$

the strong binding potential of T to M is, in effect, employed to pry A from M in AMD. These processes separately involve very large free energy changes, but in concert the net basic free energy drop (Eq. 5.38) is small (30). Thus, in the process 5.38, the high free energy of M + T relative to M*T has been largely transferred to the high free energy of A + M relative to AM (i.e., there is now the potential for strong binding of A to M in place of T to M).

The free energy changes in M*T → M**D (Fig. 5.7b) and in M**D → M‡D (Fig. 5.7c) are also small. The first of these is the actual ATP hydrolysis step. Thus, in summary so far,

$$\begin{gathered} \mathrm{AMD}\ (x \cong 0) + \mathrm{T} \rightarrow \mathrm{A} + \mathrm{M^{**}D} + \mathrm{D} \rightarrow \mathrm{A} + \mathrm{M^{\ddagger}D} + \mathrm{D} \\ A_4\ (x \cong 0) + \mu_\mathrm{T} \rightarrow A_1 + \mu_\mathrm{D} \rightarrow A_2 + \mu_\mathrm{D} \end{gathered} \tag{5.39}$$

are the first two steps in the cycle (4 → 1 → 2), and both take place with only small drops in basic free energy. The high initial free energy of M + T relative to M*T largely resides at this stage in the separation of A and M, i.e., in the free energy of A + M relative to AM.

Although the binding of A to M is very strong in AMD, as a consequence of the different conformation involved, the binding of A to M in AM‡D is weak. Thus, in the next step of the cycle (2 → 3),

$$\begin{gathered} \mathrm{A} + \mathrm{M^{\ddagger}D} + \mathrm{D} \rightarrow \mathrm{AM^{\ddagger}D}\ (x \cong a) + \mathrm{D} \\ A_2 + \mu_\mathrm{D} \rightarrow A_3\ (x \cong a) + \mu_\mathrm{D}, \end{gathered} \tag{5.40}$$

the basic free energy drop is again small. Since A is now bound again to M, the high free energy of A + M relative to AM referred to above has been transformed here to the high free energy of AM‡D ($x \cong a$) relative to AMD ($x \cong 0$). That is, the free energy is stored, so to speak, in the potential conformational change (1, 29) AM‡D → AMD.

As we shall see below, in order for muscle contraction to operate with the observed high efficiency, it is essential that the major portion of the total free energy drop $\mu_T - \mu_D$ be reserved for the next stage. In other words, it is essential that *all* of the transitions considered so far,

$$\text{AMD } (x \cong 0) \rightarrow \text{M*T} \rightarrow \text{M**D} \rightarrow \text{M}^{\ddagger}\text{D} \rightarrow \text{AM}^{\ddagger}\text{D } (x \cong a),$$

take place with only small basic free energy drops. Or, to put it another way, all of the states

$$\text{AMD } (x \cong 0) + \text{T, A} + \text{M*T} + \text{D, A} + \text{M**D} + \text{D,}$$
$$\text{A} + \text{M}^{\ddagger}\text{D} + \text{D, AM}^{\ddagger}\text{D } (x \cong a) + \text{D}$$

must be high free energy states: the high initial free energy remains "stored," in large part, in one form or another throughout the steps 4 → 1 → 2 → 3. Note that *all* of these transitions and states must have the properties just described—there is no "special" transition or state.

In the homogeneous solution case (dashed free energy levels at left of Fig. 5.7c), in the last step of the cycle (3 → 4) the large amount of free energy stored in the conformation AM‡D ($x = a$) relative to AMD ($x = 0$) will simply be dissipated as heat. The whole cycle (Eq. 5.37) will then have dissipated the total free energy drop $\mu_T - \mu_D$.

In the muscle contraction case, the transition 3 → 4 will generally occur in the neighborhood of $x = a$:

$$\begin{aligned} &\text{AM}^{\ddagger}\text{D } (x \cong a) + \text{D} \rightarrow \text{AMD } (x \cong a) + \text{D} \\ &A_3\ (x \cong a) + \mu_D \rightarrow A_4\ (x \cong a) + \mu_D. \end{aligned} \tag{5.41}$$

Once again there will be a small free energy change only (also essential for high efficiency). The conformational change AM‡D → AMD has taken place but, unlike the situation in a homogeneous solution, it is not accompanied here by a large dissipation of free energy because state 4 (AMD) *is itself elastically distorted* into a "high free energy" state at this value of x. That is, at $x \cong a$ (Fig. 5.7c), AM‡D (undistorted) and AMD (elastically distorted) have essentially the same (high) basic free energy. Thus the high or stored free energy traced above is transferred for the last time: it is now stored in the state AMD ($x \cong a$, distorted) relative to the same state AMD ($x \cong 0$, undistorted) *as elastic free energy*.

Finally, to return to the starting point AMD, $x \cong 0$, the system remains in the same attached state AMD = 4 as x changes from $x \cong a$ to $x \cong 0$. This process is unlike all the other changes mentioned above. The previous changes have all been *transitions* (changes in state), with accompanying free energy dissipation (as heat). Here the state does not change, force (dA_4/dx) is exerted by the cross-bridge on the actin filament over the entire interval in x, and, instead of more free energy dissipation, work is done by the system (on the load) in amount

$$W \cong A_4\,(x \cong a) - A_4\,(x \cong 0).$$

Roughly speaking, the free energy $A_3 - A_4$ (left side of Fig. 5.7c) dissipated in the last step $(3 \rightarrow 4)$ of the homogeneous solution cycle is here, instead, conserved as the work W. All the other transitions in the cycle, though, in muscle as well as in solution, contribute only free energy dissipation (for simplicity, we are ignoring a small amount of work accomplished by state 3 from the relatively small force dA_3/dx). Thus the efficiency in this more or less optimal cycle is roughly $\eta \cong W/(\mu_T - \mu_D)$. To maximize η, the total free energy drop in $4\ (x \cong 0) \rightarrow 1 \rightarrow 2 \rightarrow 3 \rightarrow 4\ (x \cong a)$ must be small, as has been stressed above, since this drop is dissipated as heat.

This whole process of free energy transduction, just described, is very much an indivisible "package" affair—requiring appropriate properties of *all* states and transitions in the cycle, though of course different states play different roles.

Elastic Free Energy in Closed Circular DNA

A new free energy transducing system that stores elastic free energy in a macromolecule has been discovered recently by Gellert *et al.* (31, 32). In this system, ATP chemical free energy is used, through the mediation of the enzyme DNA gyrase, to introduce negative superhelical turns into double-stranded closed circular DNA. The mechanism is not known yet.

REFERENCES

1. T. L. Hill, *Proc. Nat. Acad. Sci. U.S.* **64**, 267 (1969).
2. T. L. Hill, *Progr. Biophys. Mol. Biol.* **28**, 267 (1974).
3. T. L. Hill, *Progr. Biophys. Mol. Biol.* **29**, 105 (1975).
4. T. L. Hill, E. Eisenberg, Y. Chen, and R. J. Podolsky, *Biophys. J.* **15**, 335 (1975).
5. A. F. Huxley, *Progr. Biophys. Mol. Biol.* **7**, 255 (1957).
6. T. L. Hill, *Proc. Nat. Acad. Sci. U.S.* **61**, 98 (1968).
7. T. L. Hill and R. M. Simmons, *Proc. Nat. Acad. Sci. U.S.* **73**, 336, 2165 (1976).
8. H. E. Huxley, *Science* **164**, 1356 (1969).
9. F. D. Carlson and D. R. Wilkie, "Muscle Physiology." Prentice-Hall, Englewood Cliffs, New Jersey, 1974.

10. R. W. Lymn and E. W. Taylor, *Biochemistry* **10**, 4617 (1971).
11. T. L. Hill, "Thermodynamics of Small Systems," Part I. Benjamin, New York, 1963.
12. T. L. Hill, "Thermodynamics of Small Systems," Part II. Benjamin, New York, 1964.
13. C. J. Brokaw, *Biophys. J.* **16**, 1013, 1029 (1976).
14. H. E. Huxley and W. Brown, *J. Mol. Biol.* **30**, 383 (1967).
15. A. M. Gordon, A. F. Huxley, and F. J. Julian, *J. Physiol.* **184**, 143, 170 (1966).
16. A. F. Huxley and R. M. Simmons, *Nature* (*London*) **233**, 533 (1971).
17. N. A. Curtin, C. Gilbert, K. M. Kretzschmar, and D. R. Wilkie, *J. Physiol.* **238**, 455 (1974).
18. O. Kedem and S. R. Caplan, *Trans. Faraday Soc.* **61**, 1897 (1965).
19. S. R. Caplan, *J. Theoret. Biol.* **11**, 63 (1966).
20. Y. Chen and T. L. Hill, *Proc. Nat. Acad. Sci. U.S.* **71**, 1982 (1974).
21. T. L. Hill and Y. Chen, *Proc. Nat. Acad. Sci. U.S.* **71**, 2478 (1974).
22. Y. Chen, *J. Theoret. Biol.* **51**, 419 (1975).
23. T. L. Hill, "Thermodynamics for Chemists and Biologists," Chapter 7. Addison-Wesley, Reading, Massachusetts, 1968.
24. E. Eisenberg and T. L. Hill, *Progr. Biophys. Mol. Biol.* (in press).
25. T. L. Hill and E. Eisenberg, *Biochemistry* **15**, 1629 (1976).
26. T. L. Hill, *Proc. Nat. Acad. Sci. U.S.* **72**, 4918 (1975).
27. T. L. Hill, *Proc. Nat. Acad. Sci. U.S.* **73**, 679 (1976).
28. R. J. Podolsky and C. Nolan, *Cold Spring Harbor Symp. Quant. Biol.* **37**, 661 (1973).
29. R. G. Wolcott and P. D. Boyer, *Biochem. Biophys. Res. Commun.* **57**, 709 (1974).
30. S. Highsmith, R. A. Mendelson, and M. F. Morales, *Proc. Nat. Acad. Sci. U.S.* **73**, 133 (1976).
31. M. Gellert, K. Mizuuchi, M. H. O'Dea, and H. A. Nash, *Proc. Nat. Acad. Sci. U.S.* **73**, 3872 (1976).
32. M. Gellert, M. H. O'Dea, T. Itoh, and J. Tomizawa, *Proc. Nat. Acad. Sci. U.S.* **73**, 4474 (1976).

Chapter 6 Stochastics and Fluctuations at Cycle and State Levels

So far in this book we have been concerned almost exclusively with the mean values of various quantities: state probabilities, transition fluxes, cycle fluxes, etc. In this chapter we look into fluctuations of several kinds, and related topics. We start at the cycle level in Section 6.1, proceed to an intermediate level, between cycle and state, in Section 6.2 (muscle contraction), and then consider fluctuations in state probabilities, and the temporal approach to steady state, in Sections 6.3 and 6.4.

The logical next step would be to treat noise theory (i.e., frequency analysis of the fluctuations) in a following chapter. In fact, the present chapter is prerequisite to a study of noise. But we omit such a chapter for two good reasons: the mathematics required is significantly more sophisticated than that used in the rest of the book; and my colleague Dr. Yi-der Chen, who has developed most of the noise theory that is pertinent (1–4), has recently written a comprehensive review of this theory (5). Therefore, readers interested in noise theory should consult Chen's paper (5) in lieu of a chapter on noise.

One of the most noticeable trends in recent biophysical research is the use of noise as well as "mean value" measurements to establish rate constants in a kinetic mechanism.

Insofar as the present chapter is concerned, Sections 6.1 and 6.3 contain simpler mathematics than Sections 6.2 and 6.4. Some readers may wish to omit the latter sections.

6.1 Stochastics of Cycle Completions

This section (6, 7) is a continuation of Section 2.2, which should be reviewed. See also the example in Table 4.6 and the discussion of this table.

Two preliminary points on notation: (a) because fluctuations are to be studied, we must now introduce a mean value bar over some quantities; and (b) because other superscripts will be needed here, to avoid undue complexity in notation we omit the steady-state superscript ∞ on $\bar{J}$, p_i, etc.

Recall that we are considering (Section 2.2), at steady state, an arbitrary diagram with cycles $\kappa = a, b, c, \ldots$. Having introduced the rate constants k_{a+}, k_{a-}, k_{b+}, etc., that govern cycle completions (over a long period of time), we can now consider $P(r_a, t)$, $P(r_b, t)$, etc., where $P(r_a, t)$ is the probability that, after a sufficiently long time t, *any single system* of the ensemble has completed r_a *net* cycles in the plus direction around cycle a, etc. Since t is large, r_a may be regarded as a continuous variable. The normalization condition is $\int P\, dr_a = 1$.

Over a long time interval, the completion of cycles of each kind ($a, b, \ldots$) may be treated as an independent, one-dimensional biased walk with, for cycle a, a transition probability k_{a+} in the plus direction and k_{a-} in the minus direction, etc. Thus, r_a is the net number of steps of the walk in the plus direction. As is well known, the (Fokker–Planck) differential equation in $P(r_a, t)$ (similarly for $b, c, \ldots$) has the form

$$\frac{\partial P}{\partial t} = \tfrac{1}{2}(k_{a+} + k_{a-})\frac{\partial^2 P}{\partial r_a^2} - (k_{a+} - k_{a-})\frac{\partial P}{\partial r_a}, \tag{6.1}$$

with the solution

$$P(r_a, t) = [2\pi\sigma_a^2(t)]^{-1/2} \exp\left\{ - \frac{[r_a - \bar{r}_a(t)]^2}{2\sigma_a^2(t)} \right\}, \tag{6.2}$$

where the mean $\bar{r}_a$ and variance σ_a^2 of the Gaussian distribution in r_a are given by

$$\begin{aligned} \bar{r}_a(t) &= (k_{a+} - k_{a-})t = \bar{J}_a t/N \\ \sigma_a^2(t) &= (k_{a+} + k_{a-})t = (\bar{J}_{a+} + \bar{J}_{a-})t/N. \end{aligned} \tag{6.3}$$

The Gaussian distribution in r_a both moves ($\bar{r}_a$) and spreads (σ_a^2) with constant velocity.

Let $n_a(t)$ be the total number of a cycles, either a_+ or a_-, completed in time t. Since the frequency of occurrence of such cycles is just $k_{a+} + k_{a-}$, we have, from Eq. 6.3b, $\sigma_a^2(t) = n_a(t)$. Thus, in terms of n_a rather than t,

$$\bar{r}_a = (k_{a+} - k_{a-})n_a/(k_{a+} + k_{a-}), \qquad \sigma_a^2 = n_a. \tag{6.4}$$

As usual, the distribution gets *relatively* sharper as t and n_a increase:

$$\sigma_a/\bar{r}_a \sim n_a^{-1/2} \sim t^{-1/2}.$$

At this point we digress to comment on our emphasis, here and in Section 2.2, on long time intervals. Suppose, in observing the stochastic behavior of a system that started originally in state j, that a cycle has just been completed and the system is in, say, state i, where it may be that $i \neq j$ (if the diagram has more than one cycle). Now the probability that the *next* cycle (starting from state i) is a_+, a_-, b_+, etc., depends on the "left over" sequence of states (≥ 1), starting with j and ending with i, that has not yet been used in a cycle (see below). If we average over j (as we would for an ensemble of systems), and use long enough times, this "memory" effect will be averaged out, leading to the mean stochastic parameters already introduced: $k_{a+}, k_{a-}, \ldots, p_{a+}, p_{a-}, \ldots, \tau$. Because of this and other complications when dealing with short time intervals, we have not written the discrete ("master equation") version of Eq. 6.1. The short time interval problem is contended with below to some extent, primarily by means of numerical examples.

When several independent random events contribute to the fluctuating quantity of interest, both mean values and variances are additive. Equations 6.3 are examples of this (a_+ and a_- are independent). Another example is the consideration of an ensemble of N independent systems rather than a single system. This, in fact, corresponds to the experimental situation. Equations 6.1 and 6.2 apply also to the ensemble, where $t \to N\ \Delta t$ and $P(r_a, \Delta t)$ is in this case the probability that r_a is the net number of a cycles in the plus direction completed in the whole ensemble in time Δt, with

$$\begin{aligned} \bar{r}_a &= N(k_{a+} - k_{a-})\ \Delta t = \bar{J}_a\ \Delta t \\ \sigma_a^2 &= N(k_{a+} + k_{a-})\ \Delta t = (\bar{J}_{a+} + \bar{J}_{a-})\ \Delta t. \end{aligned} \tag{6.5}$$

In effect, the cycle rate constants have become (for the ensemble) Nk_{a+}, Nk_{a-}, etc. Of course now $N\ \Delta t$ must be large, but Δt may be small.

Still another example of additivity occurs in relation to cases such as that treated in Section 4.3. In this model either cycle a or cycle b transports one Na^+ across a membrane. Then $r_{Na} \equiv r_a + r_b$ and Eq. 6.2 holds for $P(r_{Na}, \Delta t)$, where

$$\begin{aligned} \bar{r}_{Na} &= N(k_{a+} - k_{a-} + k_{b+} - k_{b-})\ \Delta t = (\bar{J}_a + \bar{J}_b)\ \Delta t \\ \sigma_{Na}^2 &= N(k_{a+} + k_{a-} + k_{b+} + k_{b-})\ \Delta t \\ &= (\bar{J}_{a+} + \bar{J}_{a-} + \bar{J}_{b+} + \bar{J}_{b-})\ \Delta t. \end{aligned} \tag{6.6}$$

Recall that $\bar{J}_{a+}$, etc., are all positive. Equations 6.6 refer to fluctuations in an observable (operational) quantity, the Na^+ net flux.

Implicit above, although not previously mentioned, is the fact that equations of the form 6.1 and 6.2 also apply to individual *one-way* cycles. Thus, if $P(r_{\kappa+}, t)$ is the probability of observing $r_{\kappa+}$ cycles of type $\kappa+$ after a long time t, $P(r_{\kappa+}, t)$ is a Gaussian function in $r_{\kappa+}$ with mean and variance

$$\bar{r}_{\kappa+} = \sigma_{\kappa+}^2 = k_{\kappa+}t = \bar{J}_{\kappa+}t/N, \tag{6.7}$$

etc. Similarly, fluctuations in the one-way (+) Na^+ flux, observable via tracer studies, follow from Eqs. 6.6 on putting $k_{a-} = k_{b-} = 0$.

Numerical Example

It is of some pedagogical value, at least, to verify the results in this section and in Section 2.2 by explicit Monte Carlo calculations. This requires, as an incidental benefit, a precise definition of what is meant by a "cycle completion" (a question avoided so far). In this subsection we consider the example that has been studied most extensively (7).

The diagram examined here (Fig. 6.1a) has six cycles (Fig. 6.2b), labeled somewhat unconventionally. The rate constants are included in Fig. 6.1a. There is no need to specify a particular model that leads to this diagram. For this case, ten different computer runs of 10^5 transitions each were made, starting each run (arbitrarily) in state 3 (Fig. 6.2a). Each transition from one state to another was selected by a random number generator, with relative probabilities for the final state assigned in accordance with the various outgoing rate constants from the initial state.

Suppose the initial sequence of states (Fig. 6.2a) in a run is 326545612. This we call the "actual record." It contains two kinds of repeats: "immediate" and "nonimmediate" (in the latter case—see below—a cycle is

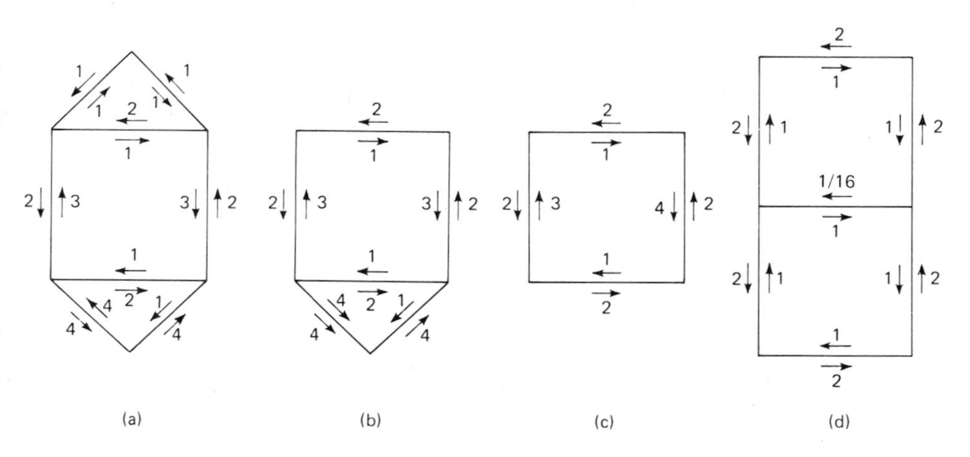

FIG. 6.1 Diagrams, with rate constants, used in Monte Carlo study of cycle fluctuations.

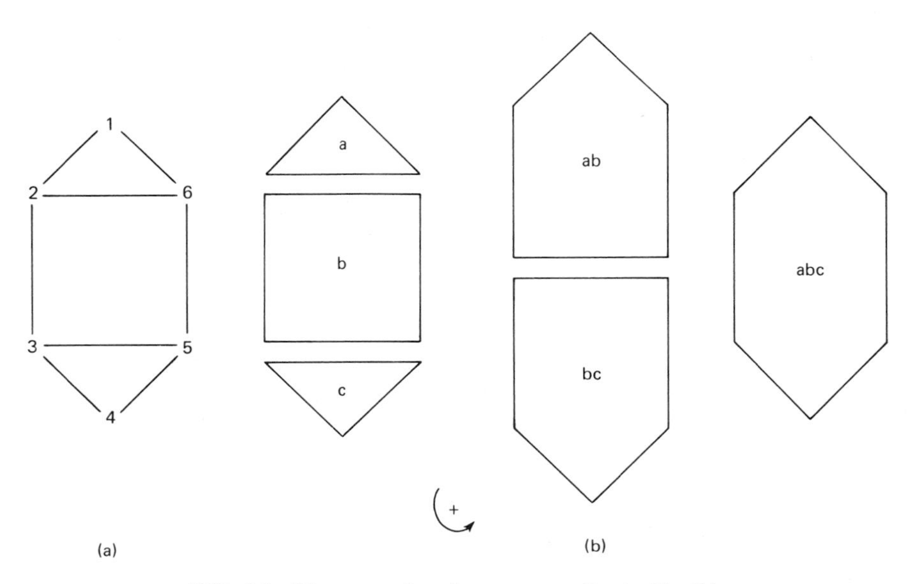

FIG. 6.2 Diagram and cycles, corresponding to Fig. 6.1a.

completed). Both kinds of repeats are canceled from the actual record, as they occur, to provide a running "effective record." Thus 326545 ("immediate" repeat of 5) becomes 3265 (effective), after cancellation of 45, and then 32656 (immediate repeat of 6) becomes 326 (effective). Finally, 32612 (nonimmediate repeat of 2) becomes 32 (effective), after cancellation of 612. Here, with the nonimmediate repeat, the cycle 2612 (i.e., $a+$ in Fig. 6.2b) has been completed. The cycle type ($a+$) is determined and tallied by the computer.

If the above actual record happened to continue with 653, we would reach another *non*immediate repeat at 32653, which would become 3 (effective) after cancellation of 2653. The second cycle completed is of type $b-$ (Fig. 6.2b).

The effective record on completion of each cycle is called a "remainder" (32 after $a+$ and 3 after $b-$, above). Note (see above) that the remainder is *not* dropped as further states in the actual record are considered in the process of determining the next cycle completion.

It is obvious that the above recipe for the determination of cycle completions records properly the actual molecular accomplishments (transport, reaction, etc.) of the system being followed (via the "actual record" of states).

We return now to the ten computer runs based on Figs. 6.1a and 6.2. The theoretical state probabilities (designated p_i^{th}) are most easily calculated in this example using the method described in Section 2.4, that is, from flux diagrams rather than from directional diagrams. These values are given in Table 6.1 along with $\tau_{tr}^{(i)}$ (see the discussion of Table 4.6). These two columns then give f_i^{th} (see Table 4.6) and also $\tau_{tr} = 0.205028$ for the mean time between transitions. The last column in Table 6.1 contains the "experimental" p_i (for the ten runs combined) based on the observed f_i (relative frequency of state i in the actual records) and $\tau_{tr}^{(i)}$. There is close agreement between p_i and p_i^{th}, as expected. Note that, in the calculation of the p_i, the experimental time is introduced via the *mean* times $\tau_{tr}^{(i)}$ only (8). To be more realistic, a distribution about each mean time should be used.

TABLE 6.1

PROPERTIES OF STATES (FIG. 6.1a)

State	p_i^{th}	$\tau_{tr}^{(i)}$	f_i^{th}	p_i
1	0.17841	1/2	0.07316	0.17858
2	0.20599	1/4	0.16893	0.20637
3	0.11463	1/9	0.21152	0.11473
4	0.08985	1/8	0.14737	0.08978
5	0.26029	1/4	0.21347	0.25984
6	0.15083	1/6	0.18555	0.15070

The second and third columns in Table 6.2 show the calculation of the twelve cycle rate constants $k_{\kappa\pm}$ from Eqs. 2.14, based on Figs. 6.1a and 6.2, where $\kappa\pm$ is the cycle index. The value of Σ is found in the p_i^{th} calculation (above) to be 4641. From Eq. 2.18b, $\tau = 4641/2152 = 2.15660$. The theoretical number of transitions per cycle completed is then $\tau/\tau_{tr} = 10.5186$ [experimental (Table 6.2) $= 10^6/95249 = 10.499$].

The last three columns in Table 6.2 refer to the ten runs *combined* (10^6 transitions). The expected (theoretical) number of cycles of each type is $\bar{r}_{\kappa\pm}^{th}$ (fourth column). These values are found as follows. The *total* expected number of cycles is $10^6\tau_{tr}/\tau = 95070.1$. The fraction of these cycles of type $a+$ is then expected to be $p_{a+}^{th} = 148/2152$ (second column), etc. That is, $\bar{r}_{a+}^{th} = 10^6\tau_{tr}p_{a+}^{th}/\tau$, etc. The *observed* number of cycles of each type is $r_{\kappa\pm}$ (fifth column). One can see at a glance that the agreement with the $\bar{r}_{\kappa\pm}^{th}$ values confirms Eqs. 2.14.

To be more quantitative about this agreement: the last column in Table 6.2 gives $\Delta_{\kappa\pm} = (r_{\kappa\pm} - \bar{r}_{\kappa\pm}^{th})/\sigma_{\kappa\pm}^{th}$, that is, the observed deviation from the expected mean relative to the standard deviation of the expected Gaussian distribution (Eq. 6.7). The mean of the 12 values of $|\Delta_{\kappa\pm}|$ is 0.617. The theoretical mean, for a very large number of values with a Gaussian distribution, is $(2/\pi)^{1/2} = 0.798$.

TABLE 6.2

PROPERTIES OF CYCLES (FIG. 6.1a)

Cycle, $\kappa_\pm$	$\Pi_{\kappa\pm}\Sigma_\kappa$	$k_{\kappa\pm}$	$\bar{r}^{\text{th}}_{\kappa\pm}$	$r_{\kappa\pm}$	$\Delta_{\kappa\pm}$
$a+$	148	0.031890	6538.3	6656	+1.456
$a-$	296	0.063779	13076.5	13019	−0.503
$b+$	256	0.055161	11309.5	11210	−0.936
$b-$	144	0.031028	6361.6	6409	+0.594
$c+$	496	0.106874	21912.1	22027	+0.776
$c-$	248	0.053437	10956.1	10880	−0.727
$ab+$	64	0.013790	2827.3	2833	+0.107
$ab-$	72	0.015514	3180.8	3195	+0.252
$bc+$	256	0.055161	11309.5	11364	+0.512
$bc-$	72	0.015514	3180.8	3235	+0.961
$abc+$	64	0.013790	2827.3	2816	−0.213
$abc-$	36	0.007757	1590.4	1605	+0.366
Total	2152	0.463695	95070.2	95249	

The above paragraph can be reinforced by calculating $\Delta_{\kappa\pm}$ for the ten *separate* runs of 10^5 transitions each. It is found in this case that the mean of the 120 values of $|\Delta_{\kappa\pm}|$ is 0.806. Furthermore, a tabulation of the 120 separate $\Delta_{\kappa\pm}$ values in intervals of 0.50 about zero produces quite respectable overall agreement with the expected Gaussian distribution.

As a further check, differences between experimental cycle numbers of each type in successive runs (1 − 2, ..., 9 − 10) were used rather than deviations from expected means, as above. That is, the 60 quantities $\delta^{12}_{a+} = (r^{(1)}_{a+} - r^{(2)}_{a+})/\sigma^{\text{th}}_{a+}$, $\delta^{34}_{a+} = (r^{(3)}_{a+} - r^{(4)}_{a+})/\sigma^{\text{th}}_{a+}$, etc., were calculated. The mean of their absolute values was found to be 1.095. The theoretical mean is $2/\pi^{1/2} = 1.128$.

Analyzed in similar fashion were various linear combinations of cycle numbers, such as $r_{a+} - r_{a-}$, $r_{a+} + r_{ab+} + r_{abc+}$, etc., also with the expected results (see Eqs. 6.3 and 6.6). We omit details since these are not independent data.

Other Examples

Similar results confirming the stochastic treatment above were found for other models (7). For the model in Fig. 6.1b (three cycles), one run of 10,000 cycle completions was made. The theoretical number of transitions per cycle completed is 11.675; observed was 11.627. The mean of the six values of $|\Delta_{\kappa\pm}|$ was found to be 0.64.

For the model in Fig. 6.1c (a single cycle), eight runs of 10^5 transitions each were made: number of transitions per cycle (theoretical) is 17.571 (!), observed (eight runs combined) was 17.594; mean of 16 values of $|\Delta_{\kappa\pm}|$ was 0.704.

For the model in Fig. 6.1d and Table 4.6 (three cycles), one run of 20,044 cycle completions was made: number of transitions per cycle (theoretical) is 12.700, observed was 12.669; mean of six values of $|\Delta_{\kappa\pm}|$ was 1.07.

Thus, all calculations made confirm Eqs. 2.14, as expected. Furthermore, they verify that, for a system observed over a long time interval, each possible kind of cycle completion ($a+$, $a-$, $b+$, etc.) can be treated as an independent random event with its own rate constant k_{a+}, k_{a-}, etc. (each k being a known function of the elementary rate constants α_{ij}). Finally, of course, these numerical results provide a check on the correctness of the recipe introduced to count cycle completions.

Analysis of Cycles for Shorter Time Intervals

We have seen that the cycle rate constants $k_{\kappa\pm}$ suffice for a stochastic analysis of cycle completions over a long time interval (large numbers of cycles). But the kinetics of individual cycle completions requires more detail. This would be necessary, for example, in the treatment of noise associated with cycle fluxes (say in active transport) over the complete frequency range. No analytical theory is available at this level. We merely present a numerical (Monte Carlo) example (7) here in the hope of stimulating future work on a proper analytical theory.

We have seen that there is a "remainder" (the effective record)—a short sequence of states—after each cycle completion. There are four essential points to be made: (a) after a cycle completion of type c' with remainder r', the probability that the next completed cycle c'' will be of any given cycle type depends on r' (but all cycle types are *possible* after any kind of remainder); (b) the mean time required for completion of c'' (counting from c') also depends on r'; (c) the probability that a given type of remainder will occur depends on the kind of cycle being completed and also on the immediately preceding remainder (in fact, each cycle type permits of only certain remainders); and (d) an average over all possible starting states (i.e., over all states in the diagram) is essential because each starting state is necessarily the starting state in all remainders that occur in a given sequence of states (actual record) and thus each starting state has its own and exclusive set of possible remainders.

To recapitulate partially: in the sequence r (remainder) c (cycle) $rcrc$ ..., each c has a "memory" of the preceding r and each r has a "memory" of the preceding rc. In contrast, the simple (long time) theory in the first part of this section includes no memory effect at all.

We turn now to an example that illustrates some but not all of the points above. We use the model in Fig. 6.1a again and arbitrarily select state 1 as the starting state. The reader can easily verify that the only possible remainders are then 1, 12, 16, 123, 165, 1265, and 1623. Also, it is easy to see that (with starting state 1): cycle types $a\pm$, $ab\pm$, and $abc\pm$ can leave only the remainder 1; $b\pm$ and $bc\pm$ can leave only 12 and 16; and $c\pm$ can leave only 123, 165, 1265, and 1623.

A single run of 20,000 cycles was made that started with state 1 and happened to end with the remainder 1265. The computer recorded (a) the numbers of completed cycles of each type that followed each kind of remainder, and (b) the number of remainders of each type that followed each kind of cycle. For simplicity in this example, the time was not considered nor did we subdivide the cycle types in (b) according to the preceding remainder (as must be done in a complete analysis).

Table 6.3 gives illustrative, partial results (six of the twelve cycle types) on (a) while Table 6.4 presents the data on (b) (omitting $a\pm$, $ab\pm$, and $abc\pm$; see above). It is evident from Table 6.3 that the relative probability of different cycle types is indeed different for different remainders. If we had subdivided cycle types according to the preceding remainder, Table 6.4 would require a third dimension.

TABLE 6.3

NUMBERS OF CYCLES FOLLOWING REMAINDERS (FIG. 6.1a)

Remainder	$a+$	$a-$	$bc+$	$bc-$	$abc+$	$abc-$
1	494	967	717	206	192	103
12	467	198	364	120	182	17
16	81	845	381	94	41	87
123	191	80	415	47	176	13
165	48	449	214	117	12	111
1265	128	8	28	69	12	1
1623	1	191	233	5	1	2
Total	1410	2738	2353	658	616	334
$\bar{r}^{\text{th}}_{\kappa\pm}$	1375	2751	2379	669	595	335

A proper theory would provide (for a given diagram and for each starting state), as functions of the α_{ij}, the *probability* of each type of remainder following each kind of cycle and preceding remainder and the *rate constant* $k^{(\text{rem})}_{\kappa\pm}$ for each type of cycle following each kind of remainder. With these available, the elementary transitions $i \rightarrow j$ could be bypassed in following the stochastics of individual cycle completions. The $k_{\kappa\pm}$ used above are, in the most general case, averages of $k^{(\text{rem})}_{\kappa\pm}$ over different starting states and over different remainders for each starting state.

TABLE 6.4

NUMBERS OF REMAINDERS FOLLOWING CYCLES (FIG. 6.1a)

Remainder	$b+$	$b-$	$bc+$	$bc-$	$c+$	$c-$	Total
12	1142	670	1154	331	—	—	3297
16	1198	736	1199	327	—	—	3460
123	—	—	—	—	1742	865	2607
165	—	—	—	—	1864	860	2724
1265	—	—	—	—	416	243	659
1623	—	—	—	—	577	293	870
Total	2340	1406	2353	658	4599	2261	
$\bar{r}^{\text{th}}_{\kappa\pm}$	2379	1338	2379	669	4610	2305	

Free Energy Bookkeeping

In Chapter 4, we encountered several examples in which a single set of basic free energy levels, repeated indefinitely, could be used to follow the "energetic" state of a single system stochastically. That is, the system performs a biased random walk on the basic free energy levels with transition probabilities α_{ij}. If less detail than this about the energetic state of the system will suffice, only one free energy level per set could be used, with these levels separated by the free energy (thermodynamic force) X. The new levels would be labeled according to the value of r (see above). Note that there would be no distinction between basic and gross free energy levels here.

As mentioned following Eq. 4.66, if there is only a *single cycle* in the diagram, say cycle a, then X might be a sum of two or more forces. Also, for this case, using the "new" less detailed set of levels, the transition probabilities between levels (over a long time period) are k_{a+} (down) and k_{a-} (up)—i.e., composites of the α_{ij} (see above). The system may now be considered to perform a biased random walk on the new levels, governed by k_{a+} and k_{a-}. If the system starts at $r = 0$, and is on some lower level $r > 0$ at some later time, the free energy loss at this point will be rX.

On the other hand, if there is only *one thermodynamic force* X in the diagram but, say, two cycles a and b that contain this force (e.g., Fig. 4.11a), a single set of r levels separated by a free energy X would again be adequate, but the transition probabilities between these levels will be (Eq. 6.6a)

$$k_+ \text{ (down)} = k_{a+} + k_{b+}, \qquad k_- \text{ (up)} = k_{a-} + k_{b-}. \tag{6.8}$$

Again the free energy loss at r (having started at $r = 0$) is rX.

The above are examples of "one-dimensional" free energy "bookkeeping" at the cycle level. However, if the diagram has more than one thermodynamic force (and flux) and more than one cycle, the number of

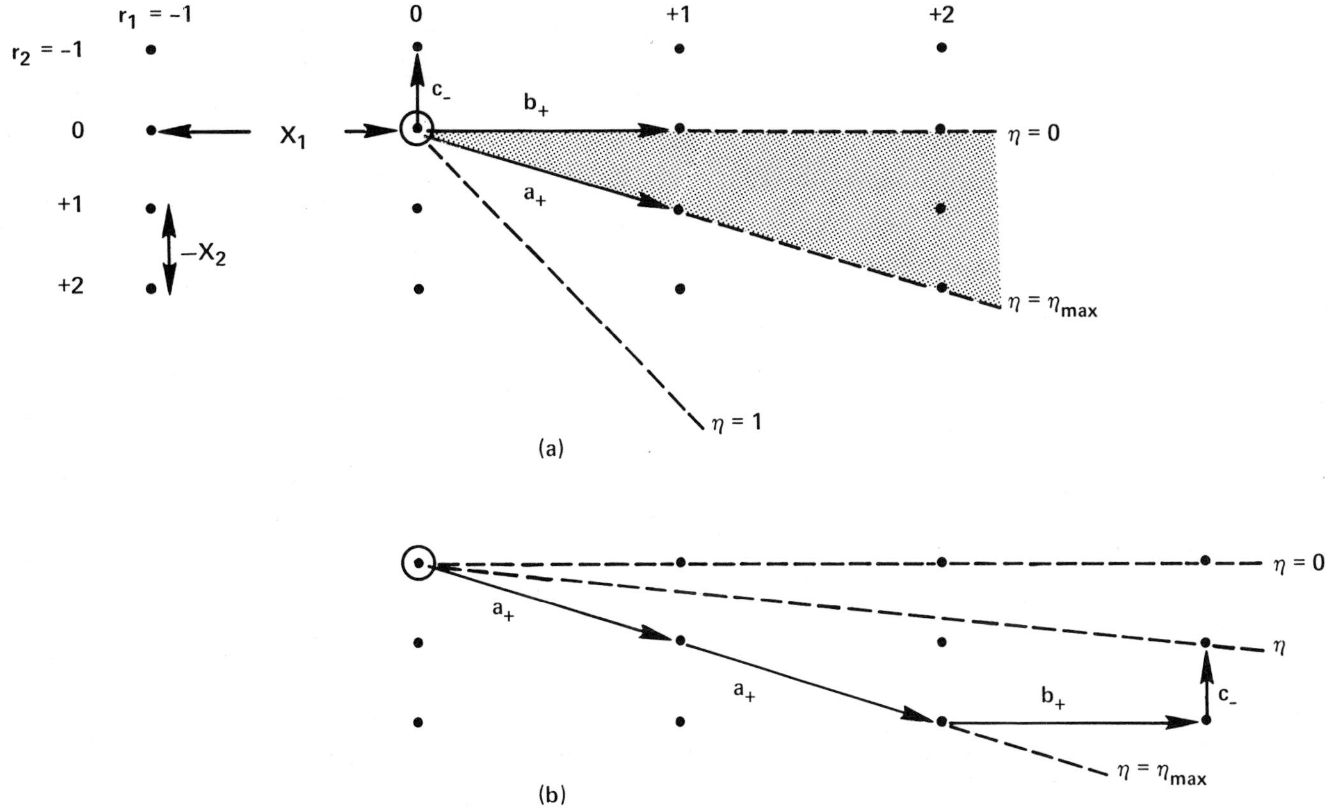

FIG. 6.3 (a) Lattice for free-energy and transport bookkeeping during stochastic behavior of system in Figs. 2.11 and 2.12 with $Na^+ = 1$, $L = 2$. See text for further details. See also Fig. 4.10 (where $-X_2/X_1 = \frac{1}{7}$). (b) Example showing efficiency η ($\eta_{max} = 2.5/8.5 = 0.294$, $\eta = \eta_{max}/3 = 0.098$).

"dimensions" needed for this purpose will be equal to the number of forces.

We consider just one example (Section 4.3; Figs. 2.11, 2.12, and 4.10), with two forces (X_1, X_2) and three cycles (a, b, c). Recall that $1 = Na^+$, $2 = L$, $X_1 > 0$, $X_2 < 0$, $X_1 > -X_2$, and out → in corresponds to positive flux in 1 or 2. We need two dimensions and we must follow the three cycles separately. We therefore use a two-dimensional free energy lattice, as in Fig. 6.3a. Each point in the lattice (r_1, r_2) represents a *net* of r_2 molecules of L and r_1 sodium ions transported in the direction out → in, starting, say, from $r_1 = r_2 = 0$. The horizontal free energy unit (double-arrow) is X_1 while the vertical unit is $-X_2$. The ratio used for $-X_2/X_1$ in Fig. 6.3a is 2.5/8.5 (in Fig. 4.10 it is $\frac{1}{7}$). The three arrows emanating from the origin show, as illustrations, the lattice displacements resulting from single cycles $a+$, $b+$, and $c-$. Stochastically, the system "walks" on the lattice points of Fig. 6.3a with transition probabilities $k_{a\pm}$, $k_{b\pm}$, and $k_{c\pm}$. At each point of such a walk, there are six possible transitions ($a\pm$, $b\pm$, $c\pm$). But over a long period of time, we will have $r_1 \sim \bar{J}_1$ and $r_2 \sim \bar{J}_2$. Also, of course, the free energy consumed is $r_1 X_1$ while the free energy recovered is $r_2(-X_2)$, with an efficiency given by Eq. 4.66a. The dashed lines in Fig. 6.3a indicate limiting values of η: $\eta = 1$ (at 45°); $\eta = \eta_{max}$ (complete coupling, with cycle a only); and $\eta = 0$ (cycle b only). The line representing the real efficiency will lie between $\eta = 0$ and $\eta = \eta_{max}$ (shaded region). Figure 6.3b illustrates this for a hypothetical case in which the net cycle fluxes occur in the ratios $\bar{J}_a : \bar{J}_b : \bar{J}_c = 2 : 1 : -1$.

6.2 Some Further Stochastic Considerations in Muscle Contraction

We have already introduced (Section 5.1) the stochastic point of view as applied to *steady isotonic contractions*. We expand on this subject here (9). But the treatment is introductory and its purpose is only to illustrate some of the possibilities. We begin with a rather general diagram and then specialize to a two-state cycle in order to make a little more explicit progress.

General Diagram

We use Fig. 6.4a, as has been our custom (see Fig. 5.2), as a sufficiently general diagram for illustrative purposes. Figure 6.4b shows schematic basic free energy levels of the states. Transitions $1 \rightleftarrows 2$ and $4 \rightleftarrows 5$ involve adsorption or desorption of ATP. These are somewhat arbitrarily singled out as "special" transitions, for counting purposes, with free energy changes shown by the *slanting* lines. The vertical lines (at $r = -1$) show the other possible

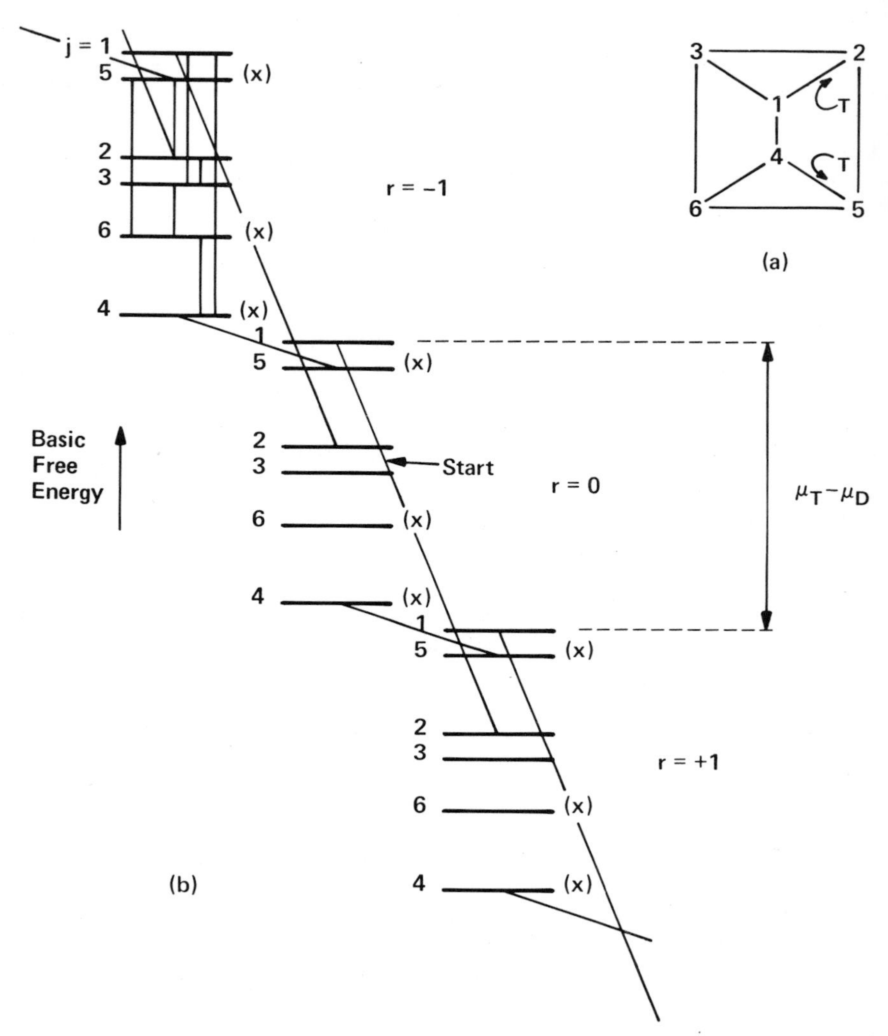

FIG. 6.4 Hypothetical sets of basic free energy levels of states in (b) corresponding to diagram in (a), for some particular value of x. Vertical lines are possible transitions (either direction); slanting lines are additional transitions for ATP adsorption–desorption. See text.

transitions and free energy changes (imagine them repeated for all values of r). The free energies of states 4, 5, and 6 are actually x-dependent (as in Fig. 5.5), but we omit this feature from Fig. 6.4b for simplicity.

The index r, here, counts the net number of ATP molecules adsorbed by a cross-bridge. In an isotonic contraction at velocity v, we always start (at $x = +d/2$) the "pass" of the nearest actin site by the cross-bridge in question at the reference value $r = 0$. The starting biochemical state is $j = 1$, 2, or 3 (unattached), usually $j = 2$ or 3 (see Fig. 6.4b); the same is true of the ending state (at $x = -d/2$). The value of r at the end of the pass may be 0, ± 1, ± 2, ..., but usually $r > 0$. Figure 6.4b should be imagined extended to include all integral values of r.

The diagram (Fig. 6.4a) provides the transition probabilities, including the x-dependence of those that involve attached states. The most detailed stochastic analysis of this system would employ the Monte Carlo method and a random number generator (8) to simulate the unpredictability of the sequence of transitions, during the pass, beginning in state 1, 2, or 3 at $x = +d/2$ and $t = 0$. In the very small interval $\Delta x = -\Delta t/v$ ($\Delta t > 0$) at any x, a transition may or may not occur, and if it does, there are always three possible transitions (see Fig. 6.4a) with relative probabilities determined by the rate constants (at x). This procedure (which we shall not elaborate on any further here) would provide a record of each transition, and where it occurred, during a pass. Probably thousands of passes (for a given v) would have to be simulated in order to form reliable averages. The starting state (at $x = +d/2$) of a given pass would be chosen as the ending state (1, 2, or 3) of the preceding pass.

Any quantity of experimental interest could obviously be computed from records of the above type. The following are examples: the probability $P_{jr}(x)$ of the cross-bridge being in biochemical state j with index r at x; the probability $p_j(x) = \sum_r P_{jr}(x)$ of state j at x, irrespective of r (this is the state probability function already used extensively in Chapter 5); the probability $P_r(x) = \sum_j P_{jr}(x)$ of index r at x, irrespective of j; the probability $P_r(-d/2)$ of index r at the end of a pass; the mean value of r, $\bar{r}(-d/2)$, at the end of a pass (obviously closely related to the ATP flux); the work done in each pass (for example, see $a + b + c$ in Fig. 5.5); the mean work and force per pass ($\bar{W} = \bar{F}d$); etc.

If a sufficient number of simulations are carried out, this method will necessarily give essentially the same results (10) as the differential equation (in the p_j) method previously used in Section 5.2, plus many further details. If the diagram is complicated enough, the Monte Carlo method might become the more practical computational procedure (instead of solving a battery of simultaneous differential equations).

A level of detail *intermediate* between the Monte Carlo method and differential equations in the p_j is an analytical stochastic approach based on differential equations in the P_{jr} (above). This section is devoted primarily to this approach.

In effect, for this purpose, we make a further subdivision of all cross-bridges in biochemical state j at x according to their r values. The six states j in Fig. 6.4a become (in principle) an infinite number of substates jr in Fig. 6.4b. Just as the differential equations in the p_j can be written merely by inspection of Fig. 6.4a, those in the P_{jr} follow in the same way from Fig. 6.4b. The rate constants are the same in the two cases. For example, for $j = 1$ and 3,

$$dP_{1r}/dt = -v\, dP_{1r}/dx = \alpha_{21} P_{2,r+1} + \alpha_{31} P_{3r} + \alpha_{41}(x)P_{4r} - [\alpha_{12} + \alpha_{13} + \alpha_{14}(x)]P_{1r} \quad (r = 0, \pm 1, \ldots). \tag{6.9}$$

$$dP_{3r}/dt = -v\, dP_{3r}/dx = \alpha_{23} P_{2r} + \alpha_{13} P_{1r} + \alpha_{63}(x)P_{6r} - [\alpha_{32} + \alpha_{31} + \alpha_{36}(x)]P_{3r} \quad (r = 0, \pm 1, \ldots). \tag{6.10}$$

Summation over r gives the corresponding differential equations in p_1 and p_3 (Eq. 5.11).

The normalization conditions are

$$\sum_j \sum_r P_{jr}(x) = 1 = \sum_j p_j(x) = \sum_r P_r(x). \tag{6.11}$$

The mean force $\bar{F}$ is still most simply calculated from Eq. 5.9. But the ATP flux $\bar{J}$ is most easily found, in this formulation, from $\bar{J} = (v/d)\bar{r}(-d/2)$. This follows because d/v is the duration of a pass and $\bar{r}(-d/2)$ is the mean net number of ATP molecules bound to the cross-bridge per pass. Of course, by definition,

$$\bar{r}(x) = \sum_r rP_r(x). \tag{6.12}$$

The probability distribution $P_r(-d/2)$, that is, the probability of a net number of exactly r ATP molecules being bound in a pass, is of particular interest. A stochastic treatment can provide this distribution; without it (Section 5.2), we know only the mean of the distribution (from $\bar{J}$).

Use of only the four pairs of "slanting" (ATP adsorption–desorption) transitions for each r, in Fig. 6.4b, gives the differential equation in P_r:

$$\begin{aligned} dP_r/dt = -v\, dP_r/dx &= (\alpha_{21} P_{2,r+1} - \alpha_{12} P_{1r}) \\ &+ (\alpha_{12} P_{1,r-1} - \alpha_{21} P_{2r}) \\ &+ [\alpha_{54}(x)P_{5,r+1} - \alpha_{45}(x)P_{4r}] \\ &+ [\alpha_{45}(x)P_{4,r-1} - \alpha_{54}(x)P_{5r}] \\ &(r = 0, \pm 1, \ldots). \end{aligned} \tag{6.13}$$

Note that this is a "mixed" equation in P_r and some of the P_{jr}. If we multiply both sides of this equation by $(r/d)\, dx$, integrate over x from $-d/2$ to $+d/2$, sum over r, and use $P_r(d/2) = 0$ $(r \neq 0)$, then after cancellations the right-hand side becomes $\bar{J}$ (Eq. 5.17) and the left-hand side is $(v/d)\bar{r}(-d/2)$. This verifies the equivalence of the two methods of calculating the flux.

Solution of the infinite set of differential equations in the P_{jr} is, of course, generally hopeless for a diagram as complicated as Fig. 6.4a. Actually, the use of the word "infinite" is appropriate only in the limit $v \to 0$. At $t = 0$, the probabilities P_{jr} are nonzero only for $r = 0$ and $j = 1, 2, 3$. For $t > 0$, the probability distribution spreads over other values of r (and j), especially $r > 0$. For finite v, at the end of the pass ($t = d/v$), the spread over r is effectively finite (i.e., there is convergence in the P_r). The set of differential equations in the P_{jr} that requires solution would include only the final spread in r values (this depends on v and on the degree of accuracy desired).

We turn now to the much simpler case of a two-state cycle.

Two-State Cycle

In this subsection we write out a few of the basic equations for the special case shown in Fig. 6.5a. Figure 6.5b shows the corresponding schematic system of basic free energy levels; Fig. 6.5c is the case with f' and g' omitted.

We now have $j = 1, 2$ and $r = 0, \pm 1, \pm 2, \ldots$ (but negative values of r are not possible if f' and g' are omitted). Starting at $j = 1$ and $r = 0$ (i.e., $P_{10} = 1$

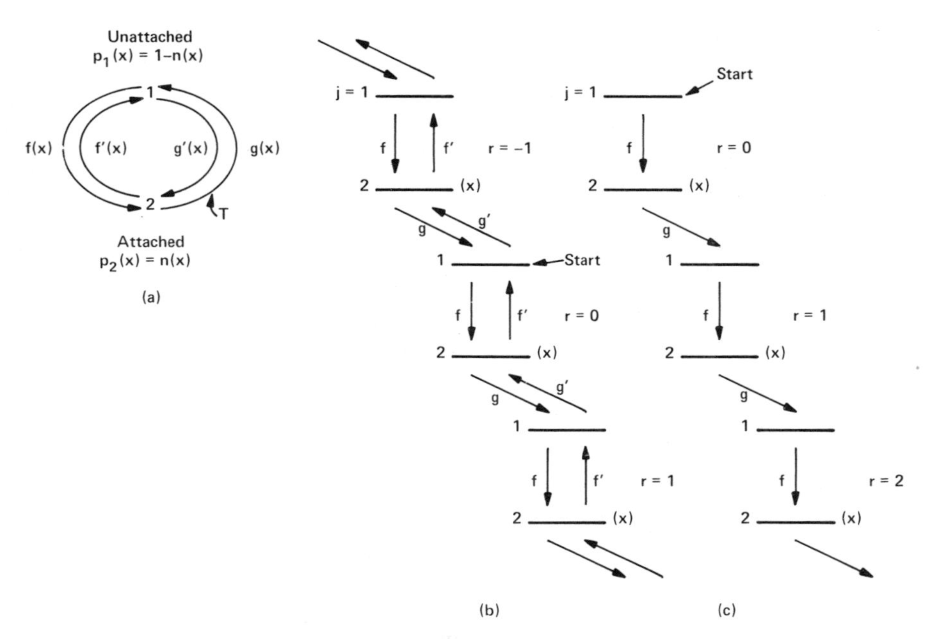

FIG. 6.5 (a) Two-state diagram. (b) Corresponding sets of free energy levels, for some x. (c) One-way cycle case. See text.

at $t = 0$), there is a "walk" on the levels and with the transition probabilities shown in Fig. 6.5b over a period of time $t = d/v$. Of course, during this period, x changes at a steady rate (from $+d/2$ to $-d/2$), and the transition probabilities are functions of x. The general trend of the walk is downhill; in Fig. 6.5c, every step is downhill.

The basic differential equations governing the stochastic walk are

$$-v\frac{dP_{1r}}{dx} = gP_{2,r-1} + f'P_{2r} - (f+g')P_{1r}$$
$$(r = 0, \pm 1, \ldots) \quad (6.14)$$
$$-v\frac{dP_{2r}}{dx} = fP_{1r} + g'P_{1,r+1} - (f'+g)P_{2r}.$$

This is still a formidable set of equations to solve. We also have

$$P_r = P_{1r} + P_{2r}, \qquad p_1 = 1 - n = \sum_r P_{1r}, \qquad p_2 = n = \sum_r P_{2r}. \quad (6.15)$$

Summation of either of Eqs. 6.14 over r gives the familiar differential equation in n, for two-state muscle models,

$$-v\,dn/dx = (f+g')(1-n) - (f'+g)n. \quad (6.16)$$

Addition of the pair of Eqs. 6.14 results in

$$-v\frac{dP_r}{dx} = gP_{2,r-1} + g'P_{1,r+1} - g'P_{1r} - gP_{2r} \qquad (r = 0, \pm 1, \ldots). \quad (6.17)$$

This determines how the probability distribution in r changes with x. The mean of this distribution is $\bar{r}$ and we denote its variance by σ^2. If we multiply Eq. 6.17 by r and then sum over r, we find

$$d\bar{r}/dt = -v\,d\bar{r}/dx = gp_2 - g'p_1. \quad (6.18)$$

The right-hand side is the flux integrand, as expected. Integration over x from $+d/2$ to x gives an expression for $\bar{r}(x)$, since $\bar{r}(d/2) = 0$. An analogous equation in $\sigma^2 = \overline{r^2} - (\bar{r})^2$ is found after multiplication of Eq. 6.17 by $(r - \bar{r})^2$ and summation over r:

$$\frac{d\sigma^2}{dt} = -v\frac{d\sigma^2}{dx} = gp_2 + g'p_1 + 2g\sum_r (r-\bar{r})P_{2r} - 2g'\sum_r (r-\bar{r})P_{1r}. \quad (6.19)$$

Here also $\sigma^2(d/2) = 0$. Thus the distribution in r starts as $P_0 = 1$ at $t = 0$ and $x = d/2$; its mean $\bar{r}$ increases as x decreases, according to Eq. 6.18; and the distribution spreads (as measured by σ^2) as x decreases, according to Eq. 6.19. We shall return to this topic below.

No numerical examples on stochastics are available. But Hill *et al.* (11) did solve Eq. 6.16 and calculate the flux (among other things) for several self-consistent choices of f, g, f', g'. Values of $\bar{r}(-d/2)$ can then be found from $\bar{J}d/v$. This is the mean net number of ATP molecules bound (or hydrolyzed) per pass. From curves (11) of $\bar{r}(-d/2)$ as a function of v, it is possible to obtain the value, call it $\bar{r}^*$, of $\bar{r}(-d/2)$ at the velocity v^* of maximum efficiency η^*. Pairs of values found for $\bar{r}^*$ and η^*, respectively, in various rather realistic (as two-state models go) cases, are: 3.4, 0.12; 2.2, 0.22; 1.5, 0.34; and 1.6; 0.37. Note that as the efficiency η^* approaches the experimental range (12) 0.40–0.50, $\bar{r}^*$ becomes not much greater than unity: high efficiencies, in two-state models, at least, are best achieved by using only one, or possibly two, ATP molecules per pass.

In the special case of a *one-way cycle* (Fig. 6.5c), $r = 0, 1, 2, \ldots$. Since $P_{10} = 1$ at $x = d/2$, we can now solve Eqs. 6.14 in succession (usually numerically). The first equation is

$$-v\, dP_{10}/dx = -f(x)P_{10}, \tag{6.20}$$

with the solution

$$P_{10}(x) = \exp\left[-\frac{1}{v}\int_x^{d/2} f(x')\, dx'\right]. \tag{6.21}$$

Next, we substitute this result in

$$-v\, dP_{20}/dx = f(x)P_{10} - g(x)P_{20} \tag{6.22}$$

and use $P_{20} = 0$ at $x = d/2$ as the boundary condition. The solution of Eq. 6.22 is then routine. It is put in the P_{11} equation, etc.

Explicit expressions for these solutions, beyond Eq. 6.21, are

$$P_{1r}(x) = \int_x^{d/2} P_{2,r-1}(x') \exp\left[-\frac{1}{v}\int_x^{x'} f(y)\, dy\right]\frac{g(x')}{v}\, dx' \qquad (r = 1, 2, \ldots), \tag{6.23}$$

$$P_{2r}(x) = \int_x^{d/2} P_{1r}(x') \exp\left[-\frac{1}{v}\int_x^{x'} g(y)\, dy\right]\frac{f(x')}{v}\, dx' \qquad (r = 1, 2, \ldots). \tag{6.24}$$

These solutions have a clear physical interpretation. For example, the probability that, during a pass, a cross-bridge is in substate $1r$ at x (Eq. 6.23) is equal to the probability that it is in substate $2, r-1$ at $x' \geq x$, multiplied by the probability $g(x')\, dx'/v$ that a transition occurs in dx', multiplied by the probability exp[ ] that no further transition occurs between x' and x (compare Eq. 6.21), with this triple product summed over all x' from x to $d/2$.

In the special case

$$\begin{aligned} g &= \text{const} && -d/2 \le x \le +d/2 \\ f &= \text{const}' && 0 \le x \le h, \quad h < d/2 \\ &= 0 && \text{otherwise,} \end{aligned} \tag{6.25}$$

it is simple but tedious to obtain analytical expressions for the first five of these probability functions from Eqs. 6.23 and 6.24. But this is left to the interested reader.

We return now to the two-way, two-state cycle (Fig. 6.5a,b) but consider another special case, namely, steady isotonic contractions in the *limit of small velocities*. In this case, large values of r predominate and we can treat $P(r, x)$ as a continuous Gaussian function in r with mean $\bar{r}(x)$ and variance $\sigma^2(x)$. We assume that v is small enough so that transitions $\Delta r = \pm 1$ occur with x effectively constant.

Since r is large, we follow r values only (ignoring j), just as in Section 6.1. But the cycle rate constants are now functions of x:

$$\begin{aligned} k_+(x) &= fg/(f + f' + g + g'). \\ k_-(x) &= f'g'/(f + f' + g + g'). \end{aligned} \tag{6.26}$$

Equation 6.1 becomes here

$$\frac{\partial P}{\partial t} = -v\frac{\partial P}{\partial x} = \tfrac{1}{2}[k_+(x) + k_-(x)]\frac{\partial^2 P}{\partial r^2} - [k_+(x) - k_-(x)]\frac{\partial P}{\partial r}. \tag{6.27}$$

As can easily be verified by direct substitution, the solution of this equation is the Gaussian function in r,

$$P(r, x) = [2\pi\sigma^2(x)]^{-1/2} \exp\left\{-\frac{[r - \bar{r}(x)]^2}{2\sigma^2(x)}\right\}, \tag{6.28}$$

where

$$\bar{r}(x) = \frac{1}{v}\int_x^{d/2} [k_+(x') - k_-(x')]\, dx' \tag{6.29}$$

and

$$\sigma^2(x) = \frac{1}{v}\int_x^{d/2} [k_+(x') + k_-(x')]\, dx'. \tag{6.30}$$

Equivalent differential expressions are

$$d\bar{r}/dt = -v\, d\bar{r}/dx = k_+ - k_-, \tag{6.31}$$

$$d\sigma^2/dt = -v\, d\sigma^2/dx = k_+ + k_-. \tag{6.32}$$

Thus, $P(r, x)$ starts $(x = +d/2)$ as a δ-function in r $(\sigma^2 = 0)$ at $r = 0$, but then spreads (σ^2) and moves $(\bar{r})$ during the pass in accordance with Eqs. 6.29 and 6.30. The final distribution is that at $x = -d/2$.

Equation 6.31 agrees with Eq. 6.18 if we substitute the isometric $(v = 0)$ expressions for p_1 and p_2 in this equation. The corresponding relation between Eqs. 6.19 and 6.32 is not simple because of the "mixed" nature of Eq. 6.19 and because the difference (at small v) between the last two terms in this equation is a quantity of the same order of magnitude as the other two terms. There are similar complications in comparing Eq. 6.27 with the continuous version (r large) of Eq. 6.17.

6.3 State Stochastics: A Two-State System

This section is presented as an easy version of and introduction to the next section. We consider the very simple system shown in Fig. 6.6. This system will approach equilibrium at $t = \infty$ since there are no cycles. We examine the stochastics of state occupation numbers (or probabilities), including time dependence. A more detailed treatment of this problem will be found in Hill (13), pp. 130–138. Also Appendix 6 is closely related.

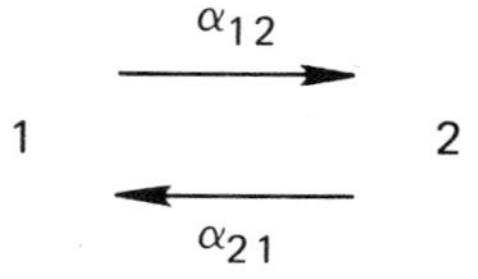

$$1 \underset{\alpha_{21}}{\overset{\alpha_{12}}{\rightleftarrows}} 2$$

FIG. 6.6 Simple two-state system used to illustrate time-dependent approach to equilibrium.

There are N systems in this ensemble, of which N_1 are in state 1 and N_2 in state 2. The probability that the ensemble, at any time t, has occupation numbers N_1, N_2 is denoted by $P(N_1, N_2, t)$, where N_1 and N_2 are discrete variables (positive integers, or zero) while t is a continuous variable. Of course only one of N_1 and N_2 is independent, since $N_1 + N_2 = N$. That is, $P(N_1, N_2, t)$ is also the probability that an ensemble of N total systems has, say, N_1 systems in state 1 at t. The N_1, N_2 notation in P is used here because of the analogy with the next section.

The so-called "master equation" for the ensemble is then

$$\begin{aligned}\frac{dP(N_1, N_2, t)}{dt} = {} & \alpha_{12}[(N_1 + 1)P(N_1 + 1, N_2 - 1, t) \\ & - N_1 P(N_1, N_2, t)] \\ & + \alpha_{21}[(N_2 + 1)P(N_1 - 1, N_2 + 1, t) \\ & - N_2 P(N_1, N_2, t)]. \end{aligned} \tag{6.33}$$

That is, the ensemble state N_1, N_2 loses probability (negative sign) by transitions

$$N_1, N_2 \begin{cases} \to N_1 - 1, N_2 + 1 & (\alpha_{12}) \\ \to N_1 + 1, N_2 - 1 & (\alpha_{21}) \end{cases}$$

and gains probability by transitions

$$\begin{matrix} (\alpha_{12}) & N_1 + 1, N_2 - 1 \\ (\alpha_{21}) & N_1 - 1, N_2 + 1 \end{matrix} \Big\} \to N_1, N_2 .$$

Equation 6.33 contains much more detail than the rate equations introduced in the earlier chapters of this book (hence, the term "master equation"). The earlier equations pertained to *mean values* of occupation numbers; here we have information about the complete distribution in occupation numbers (i.e., fluctuations).

The mean value equation in $\bar{N}_1$ follows from Eq. 6.33 on multiplication by N_1 followed by summation over N_1:

$$\begin{aligned} \sum_{N_1} N_1 \frac{dP}{dt} = \frac{d}{dt} \sum_{N_1} N_1 P = \frac{d\bar{N}_1}{dt} \\ = \alpha_{12} \sum_{N_1} (N_1 - 1) N_1 P - \alpha_{12} \sum_{N_1} N_1^2 P \\ + \alpha_{21} \sum_{N_1} (N_1 + 1) N_2 P - \alpha_{21} \sum_{N_1} N_1 N_2 P. \end{aligned} \tag{6.34}$$

Here P means $P(N_1, N_2, t)$, and the indices have been shifted where necessary to give P in each term on the right-hand side of the equation. After cancellation of the quadratic terms, we are left with

$$d\bar{N}_1/dt = -\alpha_{12}\bar{N}_1 + \alpha_{21}\bar{N}_2 . \tag{6.35}$$

If we define $p_1 = \bar{N}_1/N$ and $p_2 = \bar{N}_2/N$, then we have

$$dp_1/dt = -\alpha_{12} p_1 + \alpha_{21} p_2 . \tag{6.36}$$

This is the expected result (Fig. 6.6 and Chapter 1).

The solution of Eq. 6.35 can be put in the form

$$\bar{N}_1(t) - \bar{N}_1^e = [\bar{N}_1(0) - \bar{N}_1^e]\, e^{-(\alpha_{12}+\alpha_{21})t}, \tag{6.37}$$

where (e refers to equilibrium, as usual)

$$\begin{aligned} 0 = -\alpha_{12}\bar{N}_1^e + \alpha_{21}\bar{N}_2^e \\ \bar{N}_1^e/\bar{N}_2^e = \alpha_{21}/\alpha_{12} . \end{aligned} \tag{6.38}$$

The mean value $\bar{N}_1(t)$ approaches its final value $\bar{N}_1^e$ at a rate determined by $\alpha_{12} + \alpha_{21}$.

The explicit expression for $P(N_1, N_2)$ at $t = \infty$ is

$$P^e(N_1, N_2) = \frac{(p_1^e)^{N_1}(p_2^e)^{N_2}N!}{N_1!\,N_2\,!}. \tag{6.39}$$

This can be checked by substitution of Eq. 6.39 into Eq. 6.33, since we must have $dP/dt = 0$ at $t = \infty$. The relation (Eq. 6.38) $\alpha_{12}\, p_1^e = \alpha_{21}\, p_2^e$ is needed in carrying out this check.

The variances at any t are defined by

$$\sigma_{ij}^2 = \overline{N_i N_j} - \bar{N}_i \bar{N}_j \qquad (i, j = 1, 2). \tag{6.40}$$

In this special case there is only one independent variance (since only one of N_1 and N_2 is independent). It is easy to see that $\sigma_{11}^2 = \sigma_{22}^2 = -\sigma_{12}^2$. So we drop subscripts and write σ^2 for either σ_{11}^2 or σ_{22}^2.

By introducing a "grand partition function" (14)

$$\Xi = [1 + (p_2^e/p_1^e)]^N, \tag{6.41}$$

one can express means and variances, based on Eq. 6.39, as simple derivatives of Ξ with respect to the variable p_2^e/p_1^e. In this way we find, at equilibrium,

$$\begin{aligned}(\sigma^2)^e &= \bar{N}_1^e - [(\bar{N}_1^e)^2/N] \\ &= \bar{N}_1^e \bar{N}_2^e/N \\ &= p_1^e p_2^e N.\end{aligned} \tag{6.42}$$

Thus the probability distribution in N_1, at equilibrium, has mean $p_1^e N$ and variance $p_1^e(1 - p_1^e)N$, where (Eq. 6.38) $p_1^e = \alpha_{21}/(\alpha_{12} + \alpha_{21})$. The ratio $\sigma^e/\bar{N}_1^e$ is of order $N^{-1/2}$, as usual: if N is very large, the equilibrium distribution in N_1 is very sharp.

Corresponding to Eq. 6.35, we now derive an expression for $d\sigma^2/dt$ (at any t). We use

$$\begin{aligned}d\sigma^2/dt &= d(\overline{N_1^2} - \bar{N}_1^2)/dt \\ &= \sum_{N_1} N_1^2 \frac{dP}{dt} - 2\bar{N}_1 \frac{d\bar{N}_1}{dt},\end{aligned} \tag{6.43}$$

where the last derivative is already available from Eq. 6.35. The summation is handled as in Eq. 6.34. That is, we multiply Eq. 6.33 by N_1^2, shift indices as before, and sum over N_1. The result is

$$d\sigma^2/dt = -2(\alpha_{12} + \alpha_{21})\sigma^2 + \alpha_{12}\bar{N}_1 + \alpha_{21}\bar{N}_2\,. \tag{6.44}$$

This can be put in a more illuminating form, as follows. Define the function f as

$$f(\bar{N}^e) = \bar{N}_1^e - [(\bar{N}_1^e)^2/N]. \tag{6.45}$$

This has the functional dependence of σ^2 on $\bar{N}_1$ at $t = \infty$ (Eq. 6.42). Then, *at any time t*,

$$f(\bar{N}_1(t)) = \bar{N}_1(t) - [\bar{N}_1(t)^2/N] \tag{6.46}$$

is the value that the variance at t would have *if* it had the *equilibrium functional relation* to the actual mean value $\bar{N}_1(t)$ that exists at t. From Eqs. 6.35, 6.44, and 6.46, we find

$$\frac{d}{dt}[\sigma^2(t) - f(\bar{N}_1(t))] = -2(\alpha_{12} + \alpha_{21})[\sigma^2(t) - f(\bar{N}_1(t))], \tag{6.47}$$

which is much more symmetrical than Eq. 6.44. Then it follows that

$$\sigma^2(t) - f(\bar{N}_1(t)) = [\sigma^2(0) - f(\bar{N}_1(0))]\, e^{-2(\alpha_{12}+\alpha_{21})t}. \tag{6.48}$$

If we compare Eqs. 6.37 and 6.48 we see that, as t increases, the variance approaches its equilibrium relation to the mean at twice the rate that the mean approaches its equilibrium value. It can be shown (15) that the third and higher moments of $P(N_1, N_2, t)$ behave essentially like the variance in this respect. In other words, the probability distribution $P(N_1, N_2, t)$ approaches the equilibrium *shape* corresponding to $\bar{N}_1(t)$ twice as fast as $\bar{N}_1(t)$ approaches its equilibrium value: so to speak, the ensemble comes into *internal* equilibrium before it reaches final equilibrium.

We shall see in the next section that this same property still holds under much more general conditions: the diagram is arbitrary and the ensemble approaches steady-state rather than equilibrium at $t = \infty$.

6.4 State Stochastics: Arbitrary Diagram

Following closely the argument of the preceding section, we give in this section a condensed treatment (15, 16) of the same problem for an arbitrary diagram and for steady state (rather than equilibrium). A careful study of the generality and matrix algebra involved here has been made by Keizer (17).

We return to the use of superscript ∞ to indicate steady-state.

The systems we investigate are precisely those in Chapters 1–4 and 7 (also Chapter 5, at any x, for isometric transients). But here we examine the *approach* to steady state and, rather incidentally, *fluctuations* in occupation numbers of states *at* steady state. Elsewhere in the book the main concern has been with *mean values* of occupation numbers ($\bar{N}_i^\infty$), at steady state, and related quantities (fluxes, etc.).

An ensemble consists of N independent and equivalent systems. Each system can exist in the discrete set of states 1, 2, ..., n. In general, transitions are possible between any pair of states. Let $N_i(t)$ be the number of systems in state i at t. The $N_i(t)$ must satisfy

$$\sum_{i=1}^{n} N_i(t) = N. \tag{6.49}$$

The probability that a transition of type $i \to j$ will occur in some system of the ensemble in the interval δt at t ($\delta t \to 0$) is $\alpha_{ij} N_i(t)\, \delta t$, where α_{ij} is independent of t and of $\mathbf{N}$ (shorthand for the set $N_1, N_2, \ldots, N_n$). The α_{ij} have values such that the ensemble approaches a (stationary) steady state at $t = \infty$ (equilibrium is a special case).

Let $P(\mathbf{N}, t)$ be the probability that, at t, the ensemble is characterized by the set $\mathbf{N}$. Then the master equation is

$$dP(\mathbf{N}, t)/dt = \sum_i \sum_{j \neq i} \alpha_{ij}[(N_i + 1)P(\mathbf{N}', N_i + 1, N_j - 1, t) - N_i P(\mathbf{N}, t)], \tag{6.50}$$

where $\mathbf{N}'$ means the set $\mathbf{N}$ with omission of N_i and N_j.

An expression for $d\bar{N}_k(t)/dt$ follows, after appropriate splitting up of the double sum in Eq. 6.50:

$$\begin{aligned} d\bar{N}_k(t)/dt &= (d/dt) \sum_{\mathbf{N}} N_k P(\mathbf{N}, t) \\ &= \sum_{\mathbf{N}} N_k[dP(\mathbf{N}, t)/dt] \\ &= \sum_{i \neq k} [\alpha_{ik} \bar{N}_i(t) - \alpha_{ki} \bar{N}_k(t)]. \end{aligned} \tag{6.51}$$

At steady state, this can be written

$$0 = \sum_{i \neq k} (\alpha_{ik} p_i^\infty - \alpha_{ki} p_k^\infty), \tag{6.52}$$

where $p_i^\infty = \bar{N}_i^\infty / N$ is the probability that a given system is in state i. Equations 6.49 and 6.52 may be solved for the steady-state p's, as in Chapters 1 and 2, but this is not the point here.

We surmise that at steady state

$$P^\infty(\mathbf{N}) = \frac{(p_1^\infty)^{N_1} \cdots (p_n^\infty)^{N_n} N!}{N_1! \cdots N_n!}. \tag{6.53}$$

That this is correct may be verified by substitution of Eq. 6.53 into the right-hand side of Eq. 6.50, which must equal zero at steady state for all possible sets $\mathbf{N}$. This is, of course, a much weaker requirement than detailed balance in Eq. 6.50, at equilibrium. In the verification procedure just referred to, it is necessary to make use of Eqs. 6.52.

We define the variances at any t and for any i and j by Eq. 6.40. Then it can be shown from Eq. 6.53, using the method of Eq. 6.41, that the steady-state variances are

$$\begin{aligned}(\sigma_{ii}^2)^\infty &= \bar{N}_i^\infty - N^{-1}(\bar{N}_i^\infty)^2 \\ (\sigma_{ij}^2)^\infty &= -N^{-1}\bar{N}_i^\infty \bar{N}_j^\infty \qquad (j \neq i).\end{aligned} \tag{6.54}$$

At arbitrary t, noting Eq. 6.54, we define

$$\begin{aligned}f_{ii}(t) &= \bar{N}_i(t) - N^{-1}[\bar{N}_i(t)]^2 \\ f_{ij}(t) &= -N^{-1}\bar{N}_i(t)\bar{N}_j(t) \qquad (j \neq i).\end{aligned} \tag{6.55}$$

After considerable algebra, we then find

$$d[\sigma_{kk}^2(t) - f_{kk}(t)]/dt = 2\sum_i \alpha_{ik}[\sigma_{ik}^2(t) - f_{ik}(t)] \tag{6.56}$$

and, for $k \neq m$,

$$\begin{aligned}d[\sigma_{km}^2(t) - f_{km}(t)]/dt = &\sum_i \alpha_{ik}[\sigma_{im}^2(t) - f_{im}(t)] \\ &+ \sum_i \alpha_{im}[\sigma_{ik}^2(t) - f_{ik}(t)],\end{aligned} \tag{6.57}$$

where α_{kk} is defined here as

$$\alpha_{kk} \equiv -\sum_{i \neq k} \alpha_{ki}\,. \tag{6.58}$$

The expressions for $d\sigma_{kk}^2/dt$ and $d\sigma_{km}^2/dt$ themselves are rather more complicated than Eqs. 6.56 and 6.57, and are omitted.

With the definition 6.58, Eq. 6.51 simplifies to

$$d\bar{N}_k(t)/dt = \sum_i \alpha_{ik}\bar{N}_i(t). \tag{6.59}$$

Let us define the matrices (15)

$$\boldsymbol{\Lambda} \equiv -\boldsymbol{\alpha}^\dagger \qquad \text{and} \qquad \boldsymbol{\xi}(t) \equiv \boldsymbol{\sigma}(t) - \mathbf{f}(t), \tag{6.60}$$

where † means transpose. Then Eqs. 6.56, 6.57, and 6.59 can be written more compactly as

$$d\bar{\mathbf{N}}(t)/dt = -\boldsymbol{\Lambda}\bar{\mathbf{N}}(t) \tag{6.61}$$

and

$$d\boldsymbol{\xi}(t)/dt = -\boldsymbol{\Lambda}\boldsymbol{\xi}(t) - \boldsymbol{\xi}(t)\boldsymbol{\Lambda}^\dagger. \tag{6.62}$$

Equations 6.61 and 6.62 (approach to steady state) have exactly the same form as Eqs. 40 and 41 of Hill and Plesner (15) (approach to equilibrium).

Briefly (15), the solutions $\bar{N}_i(t)$ of Eq. 6.61 are $\bar{N}_i^\infty$ plus sums of exponentials $\exp(-\lambda_k t)$, where λ_k is the kth (nonzero and positive) eigenvalue of the matrix $\boldsymbol{\Lambda}$ (assuming no degeneracy). Also, the solutions $\xi_{ij}(t)$ of Eq. 6.62 are sums of exponentials of the form $\exp[-(\lambda_k + \lambda_l)t]$ (same eigenvalues). Thus, roughly speaking, the variance, $\boldsymbol{\sigma}^2(t)$, approaches its steady-state functional dependence $\mathbf{f}$ on the mean, $\bar{\mathbf{N}}(t)$, at twice the rate that $\bar{\mathbf{N}}(t)$ approaches its steady-state value.

We call attention, again, to the paper by Keizer (17) which, among other things, makes the preceding paragraph much more precise and general.

Open Ensembles

Throughout the book we have taken the total number of systems N in the ensemble to be a constant. That is, the ensemble has been taken as "closed" with respect to systems (macromolecules), though open with respect to ligands, etc.

But there are cases of importance in which the real ensemble is "open" to systems. For example, a system might be a macromolecular transporting unit in a membrane that can adsorb and desorb between membrane and solution, or it might be an enzyme on a surface that can adsorb and desorb between surface and solution (and is inactive in solution). In such cases (open ensembles), N fluctuates. Actually, such fluctuations have no effect on the "mean value" treatment in Chapters 1–4 and 7: we merely replace N by $\bar{N}$. But the problem of fluctuations and of the approach to steady state for open ensembles requires separate analysis. This has been carried out by Chen (2) following essentially the method of the present section.

REFERENCES

1. Y. Chen and T. L. Hill, *Biophys. J.* **13**, 1276 (1973).
2. Y. Chen. *J. Chem. Phys.* **59**, 5810 (1973).
3. Y. Chen, *J. Theoret. Biol.* **55**, 229 (1975).
4. Y. Chen, *Proc. Nat. Acad. Sci. U.S.* **72**, 3807 (1975).
5. Y. Chen, *Adv. Chem. Phys.* (in press).
6. T. L. Hill, *Biochemistry* **14**, 2127 (1975).
7. T. L. Hill and Y. Chen, *Proc. Nat. Acad. Sci. U.S.* **72**, 1291 (1975).
8. R. Gordon, *J. Chem. Phys.* **49**, 570 (1968).
9. T. L. Hill, *Progr. Biophys. Mol. Biol.* **29**, 105 (1975).
10. C. J. Brokaw, *Biophys. J.* **16**, 1013, 1029(1976).
11. T. L. Hill, E. Eisenberg, Y. Chen, and R. J. Podolsky, *Biophys. J.* **15**, 335 (1975).
12. N. A. Curtin, C. Gilbert, K. M. Kretzschmar, and D. R. Wilkie, *J. Physiol.* **238**, 455 (1974).
13. T. L. Hill, "Thermodynamics for Chemists and Biologists," Chapter 7. Addison-Wesley, Reading, Massachusetts, 1968.
14. T. L. Hill, "Statistical Thermodynamics." Addison-Wesley, Reading, Massachusetts, 1960.
15. T. L. Hill and I. W. Plesner, *J. Chem. Phys.* **43**, 267 (1965).
16. T. L. Hill, *J. Chem. Phys.* **54**, 34 (1971).
17. J. Keizer, *J. Stat. Physics* **6**, 67 (1972).

Chapter 7 Interacting Subsystems and Multienzyme Complexes

We are considering in this book ensembles of independent and equivalent macromolecular units. This chapter is no exception, but here we examine a few cases in which each unit (or system) consists of two or more distinct enzyme molecules or subunits (or subsystems). Each subunit has its own set of states, but the subunits (within a unit) interact with each other because of proximity: the kinetic properties of any one subunit depend on the states of the other subunits.

Units or systems of this type do not differ in principle from those already studied—the same methods and theorems are applicable. But they do differ in the degree of complexity of the diagram. This complexity arises because the state of a unit depends on the state of *each* of the subunits. Consequently, in this chapter, we merely introduce the subject by means of a few relatively simple examples. Certain theoretical aspects are treated in more detail elsewhere (1). Much more complicated models can be handled numerically, by computer, but it would usually be impractical to obtain general algebraic results in such cases.

Note that we are concerned here with interactions *within* a unit, and not with interactions *between* units of the ensemble. The latter is a problem of even greater complexity, which is related to the well-known Ising problem in statistical mechanics (2–5). The so-called Bragg–Williams (2, 6–8) or mean field approximation is sometimes useful in this connection at steady state (but it is a very bad approximation for transients).

Multienzyme complexes (9–11) are a special case of systems with interacting subunits. In its simplest form, the interaction in such cases might be confined to *interlocking* (or chemically coupled) transitions in which two enzymes of the complex necessarily undergo *simultaneous* transitions because a ligand, substrate, electron pair, molecular fragment, etc., is transferred *directly* from one enzyme (plus prosthetic group, usually) to the other. We shall use the term "multienzyme complex" here, in a rather narrow sense, to refer to a system that makes use of interlocking transitions (for whatever reason) between neighboring pairs of enzymes of the complex. Besides well-known biochemical cases (9–11), it seems possible that many free energy transducing complexes in membranes are also multienzyme complexes, in this sense (e.g., Na,K-ATPase, mitochondrial respiratory chain, etc.).

As we shall see, interlocking transitions simplify the diagram kinetics somewhat so we treat examples of multienzyme complexes first, in Sections 7.1 and 7.2. We then study, in Section 7.3, an example in which two enzymes (comprising one unit) interact with each other in a more general and, hence, algebraically more complicated way. A brief discussion of oxidative phosphorylation, from the point of view of Sections 7.1 and 7.2, is given in Section 7.4.

7.1 Example: Two-Enzyme Complex

We begin the discussion with Fig. 7.1a, which does *not* represent a multienzyme complex. Here a unit comprises two three-state subunits (α and β) that interact with each other. Each subunit undergoes its own cyclic steady-state activity, and has its own flux, but in general the rate constants in one subunit depend on the state of the other subunit (Section 7.3). The basic free energy drop around each cycle is equal to the corresponding thermodynamic force, X_α or X_β. These forces are determined only by bath concentrations of ligands, substrates, etc., and are therefore *independent* of the state of the opposite subunit.

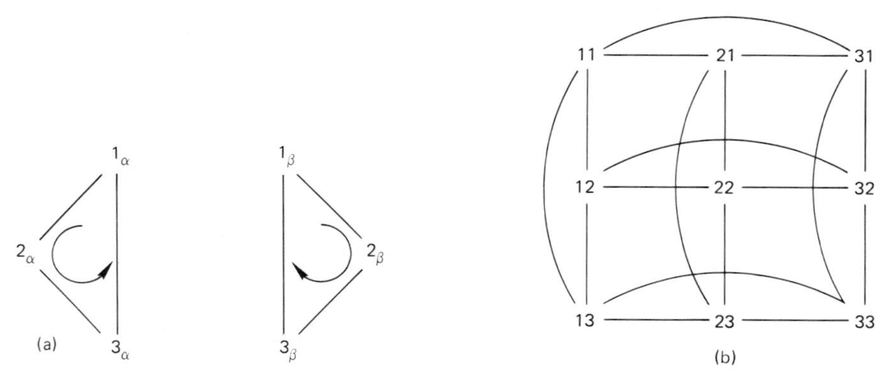

FIG. 7.1 (a) One unit consists of two three-state enzymes or subunits (α and β) that interact with each other. (b) Diagram for the system.

The state of a *unit* must be specified by *two* indices ij: i is the state of subunit α; j the state of subunit β; and $i, j = 1, 2, 3$. Figure 7.1b shows the diagram for this system. It is quite complicated. Because some cycles in Fig. 7.1b contain both forces, there will in general be some degree of thermodynamic coupling and free energy transduction (Chapter 3) between the α and β cyclic processes.

Now instead of the quite general interaction between subunits α and β that can be accommodated by Fig. 7.1, suppose α and β form a two-enzyme complex, as indicated in Fig. 7.2a. Here the transitions $1 \rightleftarrows 3$ in *both* enzymes can only occur *simultaneously* because, say, a molecule, electron pair, or molecular fragment is handed over from one enzyme to the other in the process. Ligands, substrates, etc., may enter or leave the scheme in Fig. 7.2a at any transition (Chapter 4), including $11 \rightleftarrows 33$, but this feature need not be made more explicit for present purposes. All six of the "curved" reactions ($1 \rightleftarrows 3$) in Fig. 7.1b are now eliminated and in their place we have the single process $11 \rightleftarrows 33$. The new diagram is shown in Fig. 7.2b. It is somewhat simpler than Fig. 7.1b, though still rather complicated.

Because of the interlocking reaction $11 \rightleftarrows 33$ in Fig. 7.2a, completion of α and β cycles must now go hand in hand: there is complete coupling between these two fluxes. At the same time, there can be only a single net or effective

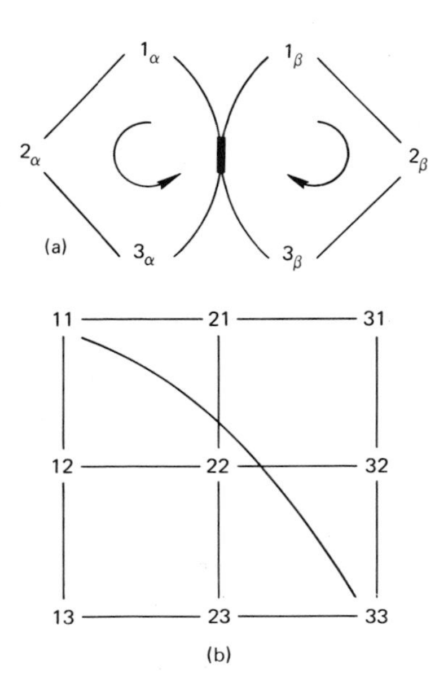

FIG. 7.2 (a) Two-enzyme complex constructed from enzymes in Fig. 7.1a, with simultaneous transition (heavy) in the two enzymes. (b) Diagram for the system.

thermodynamic force X driving the system. This resembles the behavior of a single cycle system (Chapter 3) though the present diagram has many cycles (see the beginning of Section 7.2 in this connection). If X is in fact a composite of two or more thermodynamic forces (i.e., there are two or more overall chemical or physical processes occurring in the bath or baths), free energy transduction is possible—just as in a single-cycle system with two or more forces (Chapter 3). In the well-known biochemical examples (9–11), this transduction takes the form of one chemical reaction driving another.

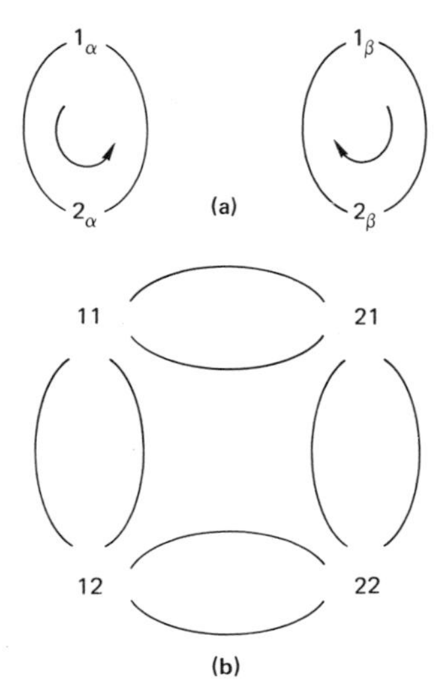

FIG. 7.3 (a) One unit consists of two two-state enzymes or subunits (α and β) that interact with each other. (b) Diagram for the system.

The Na,K-ATPase model of Stone (12) is, by our definition, a two-enzyme complex* with free energy transduction. Enzyme β hydrolyzes one ATP per cycle and, in the interlocking step, passes a phosphate group and two Na^+ (from inside) to enzyme α. The α cycle completes the overall transport of three Na^+ in $\rightarrow$ out and two K^+ out $\rightarrow$ in. The efficiency, because of the complete coupling, is given by Eq. 3.44.

Because the above example (Fig. 7.2) is rather involved, we turn to a simpler case for algebraic purposes. Figure 7.3a shows two interacting two-state subunits (compare Fig. 7.1a), with two forces X_α and X_β. The diagram is given in Fig. 7.3b. This case will be treated at length in Section 7.3. The reaction scheme for the corresponding two-enzyme complex with complete coupling and with one effective force X, is shown in Fig. 7.4a. The interlocking reaction is $11 \rightleftarrows 22$. Figure 7.4a may of course be considered a reduced

* This example was pointed out to the author by Dr. Igor Plesner.

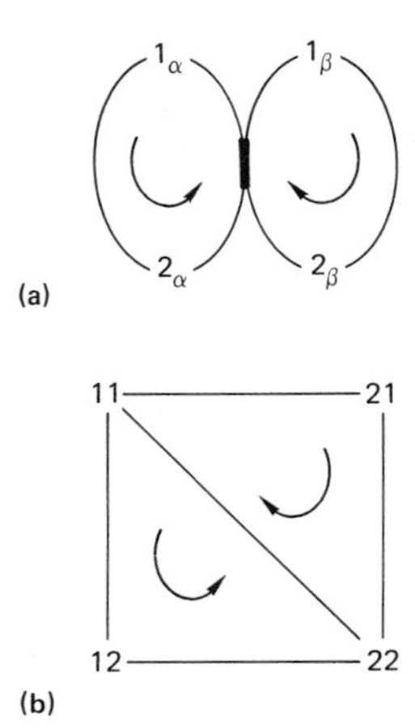

FIG. 7.4 (a) Two-enzyme complex constructed from enzymes in Fig. 7.3a, with simultaneous transition (heavy) in the two enzymes. (b) Diagram for the system.

version of Fig. 7.2a, if one state drops out of each cycle in Fig. 7.2a (Appendix 1). The diagram for Fig. 7.4a is shown in Fig. 7.4b. This should be compared with Fig. 7.2b. We now examine this case (Fig. 7.4) in some detail.

Figure 7.5a introduces the rate constant notation used for the transitions. There are two cycles, a and b (Fig. 7.5b). The large (combined) cycle has zero force and can be ignored. The same force acts in both of cycles a and b:

$$\Pi_{a+}/\Pi_{a-} = \Pi_{b+}/\Pi_{b-} = e^{X/kT}. \tag{7.1}$$

This leads to a required relation between α and β rate constants:

$$\beta_{12}\alpha^{\dagger}_{12}/\beta_{21}\alpha^{\dagger}_{21} = \alpha_{12}\beta^{\dagger}_{12}/\alpha_{21}\beta^{\dagger}_{21}. \tag{7.2}$$

The physical significance of this relation can be seen from the basic free energy levels in Fig. 7.6: the basic free energy difference between states 11 and 22 must be independent of the path. That is, Eq. 7.2 is equivalent to

$$\Delta A'_\beta + \Delta A'^{\dagger}_\alpha = \Delta A'_\alpha + \Delta A'^{\dagger}_\beta \tag{7.3}$$

where $\Delta A'_\beta \equiv A'_{11} - A'_{12}$, etc., and

$$\begin{aligned} &\beta_{12}/\beta_{21} = \exp(\Delta A'_\beta/kT), \qquad \alpha^{\dagger}_{12}/\alpha^{\dagger}_{21} = \exp(\Delta A'^{\dagger}_\alpha/kT) \\ &k_{21}/k_{12} = \exp(\Delta A'_k/kT), \end{aligned} \tag{7.4}$$

etc. It should be recalled (Chapter 4) that it is not necessary for *all* of the basic free energy steps to be downhill (as shown in the figure).

The fact that we are using two pairs of α rate constants (α and $\alpha^{\dagger}$) and two pairs of β rate constants implies that there are two kinds of interaction

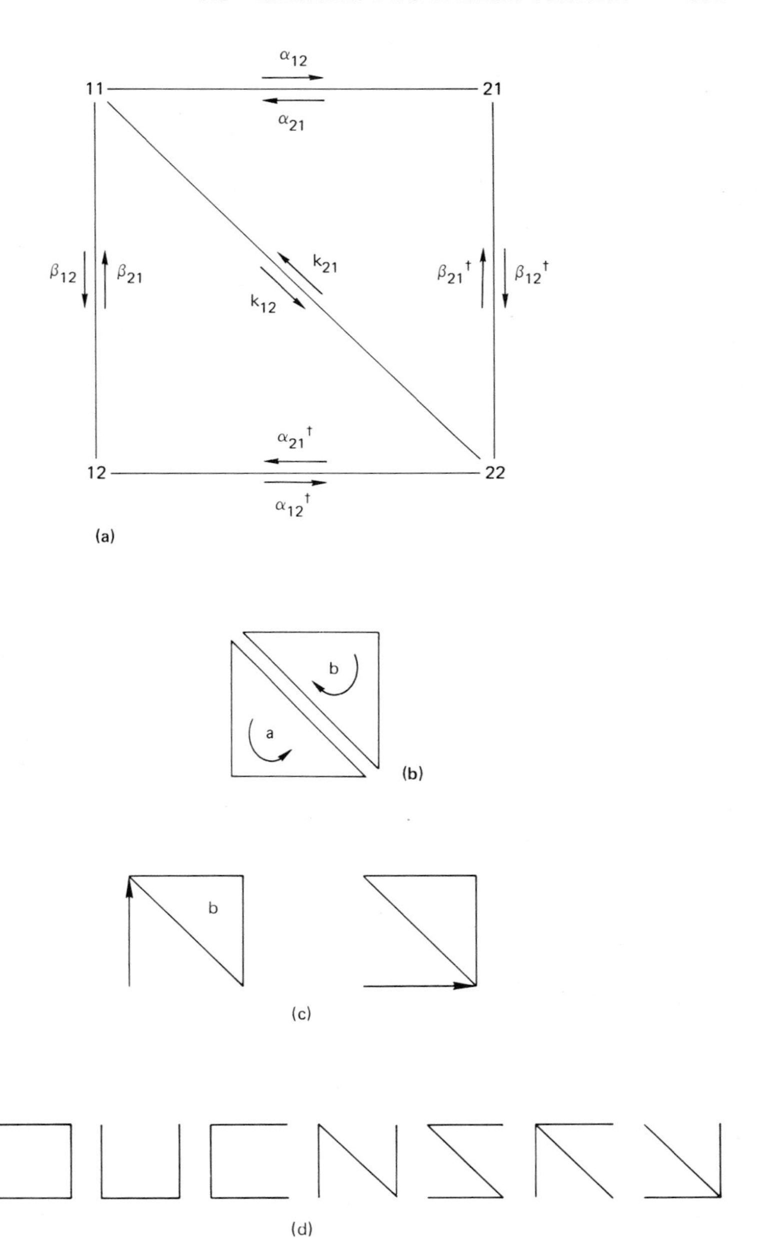

FIG. 7.5 (a) Rate constant notation for Fig. 7.4. (b) Significant cycles for the model. The combined (square) cycle has no force, no net flux, and accomplishes nothing. (c) Flux diagrams for cycle *b*. (d) Eight partial diagrams.

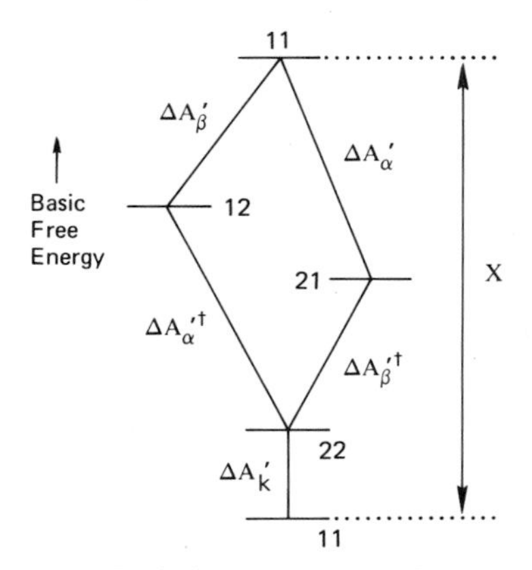

FIG. 7.6 Basic free energy levels for Fig. 7.5.

between enzymes α and β in this model: (a) there are interlocking transitions $11 \rightleftarrows 22$; and (b) in the other transitions the state of one enzyme influences the kinetics of the other. Explicitly, when α is in state 1, β has rate constants β_{12} and β_{21}, but when α is in state 2, β has rate constants $\beta_{12}^{\dagger}$ and $\beta_{21}^{\dagger}$; etc. (The dagger superscript will always be used to designate the influence of a state 2.) Equation 7.3 is interpreted further following Eq. 7.45.

If interaction (b) is not present, as could quite plausibly be the case, then

$$\begin{aligned} &\alpha_{12} = \alpha_{12}^{\dagger}, \qquad \alpha_{21} = \alpha_{21}^{\dagger}, \qquad \beta_{12} = \beta_{12}^{\dagger}, \qquad \beta_{21} = \beta_{21}^{\dagger} \\ &\Delta A'_{\alpha} = \Delta A'^{\dagger}_{\alpha}, \qquad \Delta A'_{\beta} = \Delta A'^{\dagger}_{\beta}, \qquad \Delta A'_{\alpha} + \Delta A'_{\beta} + \Delta A'_{k} = X \\ &\Pi_{a+} = \Pi_{b+} = \beta_{12}\alpha_{12}k_{21}, \qquad \Pi_{a-} = \Pi_{b-} = \beta_{21}\alpha_{21}k_{12}. \end{aligned} \tag{7.5}$$

In this event, the basic free energy differences $\Delta A'_{\alpha}$ and $\Delta A'_{\beta}$ can be attributed to the *individual* subunits. But in general, with interaction (b) present, all basic free energy differences refer to entire $\alpha\beta$ units or complexes and they cannot be decomposed into α and β contributions. See the end of Section 7.3 for illustration of this point.

Figure 7.5c shows the two flux diagrams for cycle b (cycle a is similar) while Fig. 7.5d presents the partial diagrams. From Fig. 7.5c,

$$J_a = N(\Pi_{a+} - \Pi_{a-})(\alpha_{21} + \beta_{12}^{\dagger})/\Sigma, \qquad J_b = N(\Pi_{b+} - \Pi_{b-})(\alpha_{12}^{\dagger} + \beta_{21})/\Sigma. \tag{7.6}$$

Despite Eq. 7.1, these cycle fluxes need not be equal. In fact, one of them might be negligible compared to the other (the "cycles" in Figs. 7.4a and 7.4b are not to be confused in this connection). The total flux is, of course,

$$J^{\alpha} = J_a + J_b. \tag{7.7}$$

The total rate of free energy dissipation is $J^{\infty}X$. If, say, $X = X_1 + X_2$, where some overall reaction or transport process 1 ($X_1 > 0$) drives a second reaction or process 2 ($X_2 < 0$; $X_1 > -X_2$) uphill, by means of this model (Fig. 7.4), then the efficiency of free energy transduction is $\eta = -X_2/X_1$ (compare Eq. 3.21). This is seemingly unrelated to the kinetics but, in fact, X_2, X_1, and some of the rate constants would all be functions of bath concentrations.

It is easy but tedious to find Σ and the four p_{ij}^{∞} either from the four flux diagrams (Fig. 7.5c) or from the 32 directional diagrams (Fig. 7.5d). For example, in the special case Eq. 7.5,

$$\begin{aligned}\Sigma = {} & (\alpha_{12} + \alpha_{21})(\beta_{12} + \beta_{21})(k_{12} + k_{21} + \alpha_{12} + \alpha_{21} + \beta_{12} + \beta_{21}) \\ & + k_{12}(\alpha_{21}^2 + \beta_{21}^2) + k_{21}(\beta_{12}^2 + \alpha_{12}^2) \\ & + (k_{12} + k_{21})(\beta_{12}\beta_{21} + \alpha_{12}\alpha_{21}). \end{aligned} \tag{7.8}$$

If all the *back reactions have negligible rates*, as indicated in Fig. 7.7, then the kinetic properties of the model are very simple. We have (dropping subscripts as in Fig. 7.7)

$$J_a = N\beta\alpha^{\dagger}k\beta^{\dagger}/\Sigma, \qquad J_b = N\alpha\beta^{\dagger}k\alpha^{\dagger}/\Sigma \tag{7.9}$$

$$J^{\infty} = N\alpha^{\dagger}k\beta^{\dagger}(\alpha + \beta)/\Sigma. \tag{7.10}$$

Note that the ratio J_a/J_b is equal to β/α. Also, using the method of Section 2.4,

$$\begin{aligned} \beta p_{11}^{\infty} &= \alpha^{\dagger} p_{12}^{\infty} = J_a/N = \beta\alpha^{\dagger}k\beta^{\dagger}/\Sigma \\ \alpha p_{11}^{\infty} &= \beta^{\dagger} p_{21}^{\infty} = J_b/N = \alpha\beta^{\dagger}k\alpha^{\dagger}/\Sigma \\ k p_{22}^{\infty} &= (J_a + J_b)/N. \end{aligned} \tag{7.11}$$

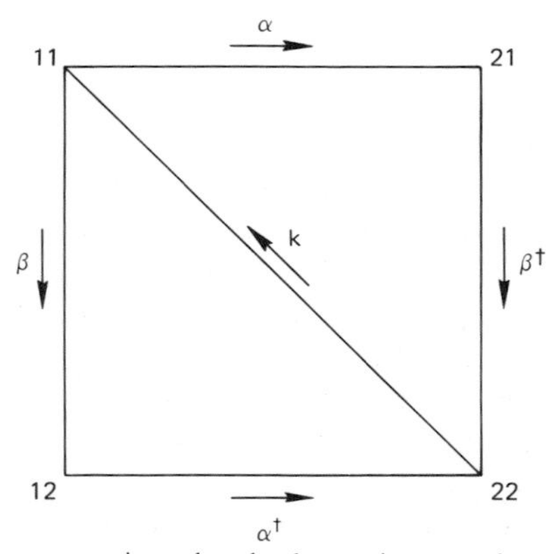

FIG. 7.7 Rate constant notation when back reactions are dropped from Fig. 7.5a.

From these we find

$$p_{11}^{\infty} = \alpha^{\dagger}k\beta^{\dagger}/\Sigma, \qquad p_{12}^{\infty} = \beta k\beta^{\dagger}/\Sigma$$
$$p_{21}^{\infty} = \alpha k\alpha^{\dagger}/\Sigma, \qquad p_{22}^{\infty} = \alpha^{\dagger}\beta^{\dagger}(\alpha + \beta)/\Sigma. \tag{7.12}$$

Normalization of the p_{ij}^{∞} then gives

$$\Sigma = k(\alpha\alpha^{\dagger} + \beta\beta^{\dagger} + \alpha^{\dagger}\beta^{\dagger}) + \alpha^{\dagger}\beta^{\dagger}(\alpha + \beta). \tag{7.13}$$

There are two classes of reaction in Figs 7.4a and 7.7, the interlocking (or coupled) reaction $22 \to 11$ and the "single" reactions (for both α and β) $1 \to 2$. After the coupled reaction occurs, there is a mean "waiting time" before the system returns to the state 22, making the coupled reaction again possible. This waiting time is taken up by the two required $1 \to 2$ transitions If there were no waiting for these single reactions, the hypothetical flux would be larger and equal to $J' = Nk$. Hence

$$\text{waiting time} = (J^{\infty}/N)^{-1} - (J'/N)^{-1}$$
$$= (\alpha\alpha^{\dagger} + \beta\beta^{\dagger} + \alpha^{\dagger}\beta^{\dagger})/\alpha^{\dagger}\beta^{\dagger}(\alpha + \beta). \tag{7.14}$$

The fraction of the total time $(J^{\infty}/N)^{-1}$ spent waiting for single reactions is

$$\text{waiting time} \times (J^{\infty}/N) = k(\alpha\alpha^{\dagger} + \beta\beta^{\dagger} + \alpha^{\dagger}\beta^{\dagger})/\Sigma. \tag{7.15}$$

The waiting time itself can be subdivided into the mean time required for the first single reaction (in either α or β) and the mean time for the second such reaction. Since, obviously (Fig. 7.7),

$$\text{first waiting time} = (\alpha + \beta)^{-1}, \tag{7.16}$$

we find, from Eq. 7.14,

$$\text{second waiting time} = (\alpha\alpha^{\dagger} + \beta\beta^{\dagger})/\alpha^{\dagger}\beta^{\dagger}(\alpha + \beta). \tag{7.17}$$

7.2 Example: Three-Enzyme Complex

One of the best known multienzyme complexes is the three-enzyme pyruvate dehydrogenase complex (9, 11) [see also Lehninger (13), pp. 450–453]. The reaction scheme, in abstract notation, is shown in Fig. 7.8a and the corresponding diagram in Fig. 7.8b. Enzymes α and β have two states each; enzyme γ has three states. The *complex* has $2 \times 3 \times 2 = 12$ states, indicated by three subscripts ijk in the order $\alpha\gamma\beta$. The arrows in Figs. 7.8a and 7.8b indicate the dominant direction for transitions. The interlocking reactions are $1_\alpha 1_\gamma \rightleftarrows 2_\alpha 2_\gamma$ and $1_\gamma 1_\beta \rightleftarrows 3_\gamma 2_\beta$.

Because of the interlocking transitions, this system (Fig. 7.8a), as well as Figs. 7.2a and 7.4a, behaves in some respects like a single-cycle system with a

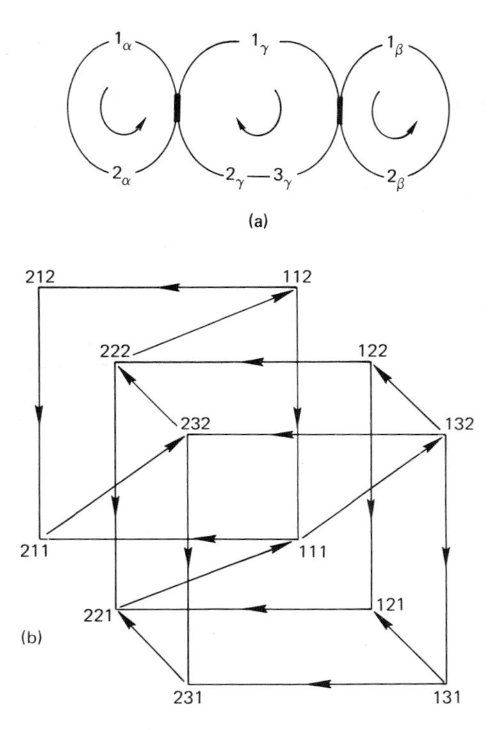

FIG. 7.8 (a) Reaction scheme, in abstract notation, for the three-enzyme pyruvate dehydrogenase complex. (b) Diagram corresponding to Fig. 7.8a. See text.

single net effective thermodynamic force X. The same would be true if Fig. 7.8a were extended linearly to include 4, 5, ... interlocking reaction loops (enzymes), and irrespective of the number of states per loop (2, 3, ...). The analogy to a single cycle is seen quite clearly by topological rearrangement of Figs. 7.2a and 7.8a. If the α loop in Fig. 7.2a is imagined folded down and under the β loop, we obtain Fig. 7.9a. If, in Fig. 7.8a, the α loop is folded *under* the γ loop while the β loop is folded *over* the γ loop (keeping both interlocking joints fixed), we obtain Fig. 7.9b. Similarly, with a four-enzyme linear complex, we have Fig. 7.9c, etc. With the reaction scheme collapsed in this way (something like a "slinky"), the reaction flux is seen not only to have the same direction in all loops but overall circulation occurs essentially *simultaneously in all loops*, kept in phase by the interlocking transitions. The net result is a "collapsed single-cycle" flux.

In order to present a not too complicated algebraic example, we now simplify Fig. 7.8 in three ways: (i) enzyme γ has two significant states, rather than three (the "cube" in Fig. 7.8b degenerates into a square); (ii) all reac-

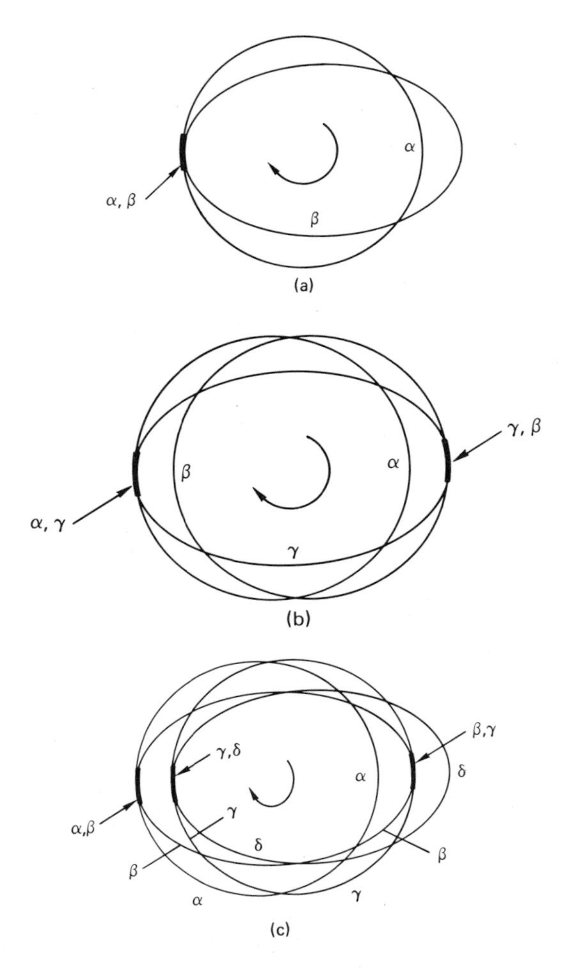

FIG. 7.9 Single-cycle aspect of linear enzyme complexes. See text. (a) Two loops or enzymes, as in Fig. 7.2a. (b) Three loops, in order α, γ, β, as in Fig. 7.8a. (c) Extension to four loops, in order α, β, γ, δ.

tions are one-way; and (iii) interactions of type (b) above are omitted* (as in Eqs. 7.5). The resulting reaction scheme and diagram, with rate constant notation, are shown in Fig. 7.10. We proceed now to deduce the steady-state kinetic properties of this system.

* This last simplification can easily be avoided if "nearest-neighbor" interactions only are included (as would be reasonable, physically). But the algebra becomes a little more complicated. See the next subsection in this connection.

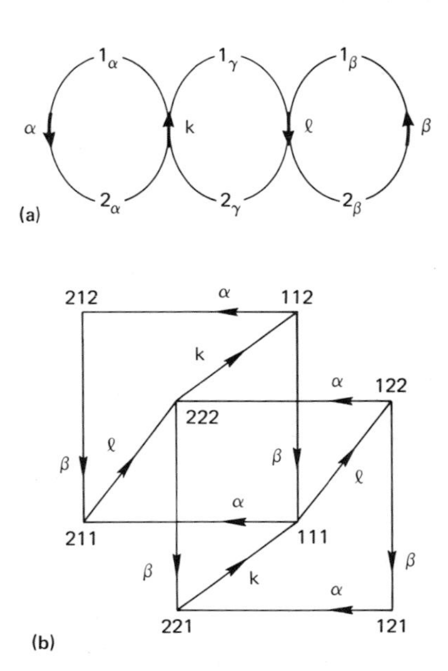

FIG. 7.10 (a) Simplified version of Fig. 7.8a. See text. (b) Diagram with one-way rate constant notation. The labeling of states is ijk in the order $\alpha\gamma\beta$, with $i, j, k = 1, 2$.

The flux diagram method of finding steady-state probabilities (Section 2.4) is very advantageous here. Figure 7.11 shows the six cycles $a, b, \ldots, f$ that have nonzero flux (if back reactions are included, the number of such cycles is twelve). All of these have the same net force X. For each cycle, the number of flux diagrams is indicated in parentheses. Since these flux diagrams are easy to enumerate, they are left to the reader (see also Eqs. 7.18). Figure 7.12 indicates which cycles contribute to each transition flux (between states).

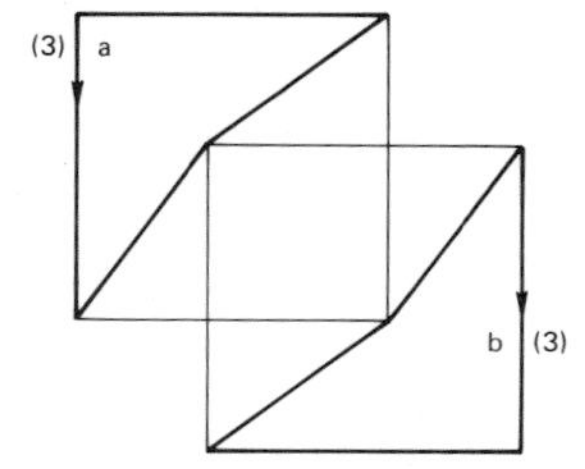

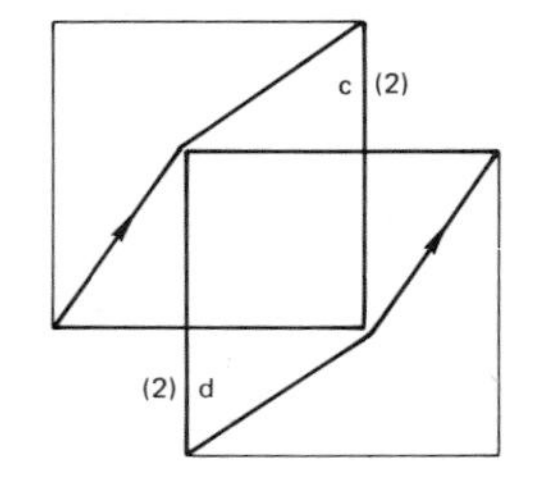

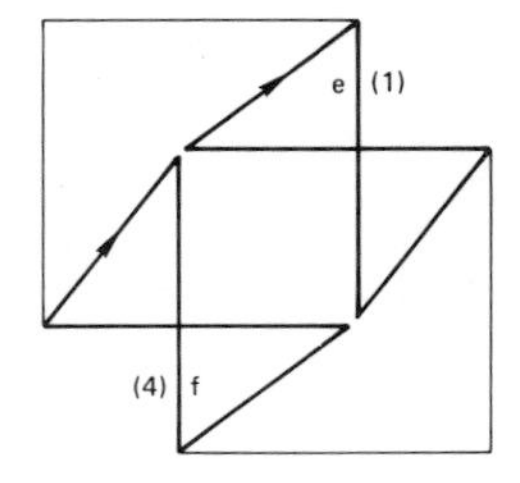

FIG. 7.11 Six cycles with nonzero flux that belong to the diagram in Fig. 7.10b. Numbers in parentheses indicate the number of flux diagrams.

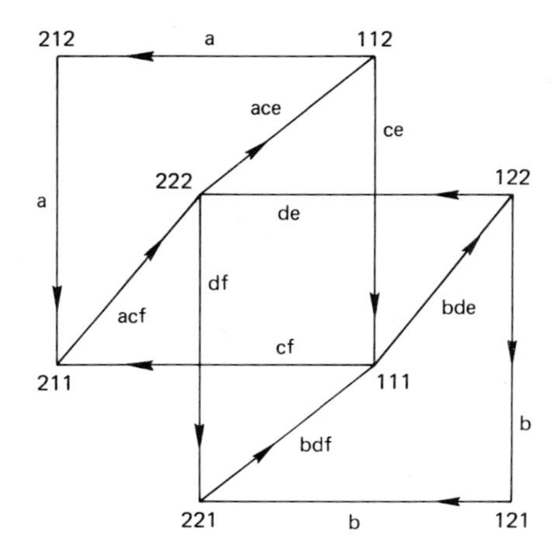

FIG. 7.12 Listing of cycles that contribute to each transition flux in Fig. 7.10b.

We find, from the flux diagrams,

$$\begin{aligned} J_a &= N\Pi_+ \alpha^2 k(\alpha + \beta + l)/\Sigma, & J_b &= N\Pi_+ \beta^2 l(\alpha + \beta + k)/\Sigma \\ J_c &= N\Pi_+ \alpha\beta k(\alpha + \beta)/\Sigma, & J_d &= N\Pi_+ \alpha\beta l(\alpha + \beta)/\Sigma \\ J_e &= N\Pi_+ \alpha\beta kl/\Sigma, & J_f &= N\Pi_+ \alpha\beta(\alpha + \beta)^2/\Sigma, \end{aligned} \tag{7.18}$$

where $\Pi_+ = \alpha\beta kl$ and Σ is yet to be determined.

The steady-state probabilities now follow immediately from Eqs. 7.18, Fig. 7.10b, and Fig. 7.12 because the (one-way) transition fluxes are very simply related to the probabilities. For example,

$$\begin{aligned} &\alpha p_{112}^{\infty} = \beta p_{212}^{\infty} = J_a/N, \qquad \beta p_{122}^{\infty} = \alpha p_{121}^{\infty} = J_b/N \\ &k p_{222}^{\infty} = (J_a + J_c + J_e)/N, \end{aligned} \tag{7.19}$$

etc. These give

$$\begin{aligned} p_{212}^{\infty} &= \alpha^3 k^2 l(\alpha + \beta + l)/\Sigma, \\ p_{121}^{\infty} &= \beta^3 k l^2(\alpha + \beta + k)/\Sigma \\ p_{222}^{\infty} &= \alpha^2 \beta kl(\alpha + \beta)(\alpha + \beta + l)/\Sigma, \\ p_{111}^{\infty} &= \alpha\beta^2 kl(\alpha + \beta)(\alpha + \beta + k)/\Sigma \\ p_{211}^{\infty} &= \alpha^2 \beta k[\alpha k(\alpha + \beta + l) + \beta(\alpha + \beta)(\alpha + \beta + k)]/\Sigma, \\ p_{112}^{\infty} &= \alpha^2 \beta k^2 l(\alpha + \beta + l)/\Sigma \\ p_{221}^{\infty} &= \alpha\beta^2 l[\beta l(\alpha + \beta + k) + \alpha(\alpha + \beta)(\alpha + \beta + l)]/\Sigma, \\ p_{122}^{\infty} &= \alpha\beta^2 k l^2(\alpha + \beta + k)/\Sigma. \end{aligned} \tag{7.20}$$

Finally, we deduce Σ by normalizing the probabilities:

$$\Sigma = \alpha^2(\alpha + \beta + l)[l(\alpha + \beta)(\beta^2 + \beta k + k^2) + \alpha\beta k^2] + \beta^2(\alpha + \beta + k)[k(\alpha + \beta)(\alpha^2 + \alpha l + l^2) + \alpha\beta l^2]. \qquad (7.21)$$

With Σ available, we now have explicit expressions for the cycle fluxes J_κ (Eqs. 7.18).

The total steady-state flux for this system, associated with the net force X, is just the sum of the six cycle fluxes:

$$J^\infty = N\Pi_+[\alpha^2 k(\alpha + \beta + l) + \beta^2 l(\alpha + \beta + k) + \alpha\beta(\alpha + \beta)(\alpha + \beta + k + l) + \alpha\beta kl]/\Sigma. \qquad (7.22)$$

To verify that this thermodynamic flux is the same as the "collapsed single-cycle" flux referred to above, we note that the *total* $1_\alpha \to 2_\alpha$ flux is the sum of the four *horizontal* transition fluxes in Fig. 7.12. This sum is seen, from Fig. 7.12, to be equal to the sum of the six cycle fluxes.

In this model, there is again (Eq. 7.14) a "waiting time" associated with the two "single" reactions at each end of the reaction scheme, Fig. 7.10a. If we let J' be the hypothetical flux without any waiting, we have

$$(J'/N)^{-1} = k^{-1} + l^{-1}$$

or

$$J' = Nkl/(k + l). \qquad (7.23)$$

This same result follows from Eq. 7.22 on letting $\alpha \to \infty$ and $\beta \to \infty$. The waiting time is then found to be

$$(J^\infty/N)^{-1} - (J'/N)^{-1} = (\alpha + \beta)[\alpha^2 k(\alpha + \beta + l) + \beta^2 l(\alpha + \beta + k)]/\alpha\beta[\ ] \qquad (7.24)$$

where [ ] is the square bracket expression in the numerator of Eq. 7.22. The fraction of the total time spent waiting is (Eq. 7.15)

$$kl(\alpha + \beta)[\alpha^2 k(\alpha + \beta + l) + \beta^2 l(\alpha + \beta + k)]/\Sigma, \qquad (7.25)$$

where Σ is given by Eq. 7.21. The terms in the numerator here can be seen to be among those in Σ, as might be expected. The same is true of Eq. 7.15.

Free Energy Levels

In the consideration of thermodynamics for this model (Fig. 7.10), it is easy to be more general. Therefore, we reintroduce backward rate constants (these are, in fact, essential for this discussion) and also interactions of type (b), above. For the latter, we make the quite realistic assumption that, in this linear array of three enzymes ($\alpha\gamma\beta$), there are nearest-neighbor interactions only (i.e., $\alpha\gamma$ and $\gamma\beta$). Thus we use rate constants $\alpha_{12}, \alpha_{21}, \beta_{12}, \beta_{21}$ if enzyme γ is in state 1 and $\alpha^{\dagger}_{12}, \alpha^{\dagger}_{21}, \beta^{\dagger}_{12}, \beta^{\dagger}_{21}$ if γ is in state 2. Also, we use l_{12}, l_{21} (α in 1), $l^{\dagger}_{12}, l^{\dagger}_{21}$ (α in 2), k_{12}, k_{21} (β in 1), and $k^{\dagger}_{12}, k^{\dagger}_{21}$ (β in 2). Figure 7.13 shows these assignments in the diagram, omitting the subscripts.

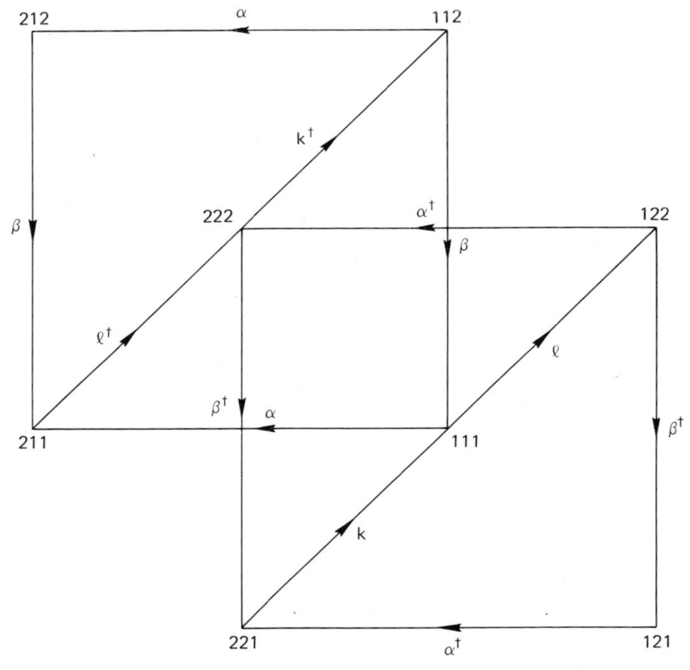

FIG. 7.13 Rate constant notation (but with subscripts omitted) when back reactions and nearest-neighbor interactions of type (b) (see text) are included in Fig. 7.10b. The arrows here indicate the dominant direction, not "one-way."

A hypothetical corresponding set of basic free energy levels is presented in Fig. 7.14. The free energy need not drop with each arrow as in this figure but the $\alpha\beta$ and $\alpha^{\dagger}\beta^{\dagger}$ parallelograms are essential (because of the nearest-neighbor assumption). If interactions of type (b) are absent, the $\alpha\beta$ and $\alpha^{\dagger}\beta^{\dagger}$ parallelograms become identical (dotted lines). The basic free energy changes corresponding to the arrows in Fig. 7.14 are related to rate constants by (Eqs. 7.4)

$$\begin{aligned} \alpha_{12}/\alpha_{21} &= \exp(\Delta A'_{\alpha}/kT), & \beta_{21}/\beta_{12} &= \exp(\Delta A'_{\beta}/kT) \\ l^{\dagger}_{12}/l^{\dagger}_{21} &= \exp(\Delta A'^{\dagger}_{l}/kT), & k_{21}/k_{12} &= \exp(\Delta A'_{k}/kT), \end{aligned} \tag{7.26}$$

etc.

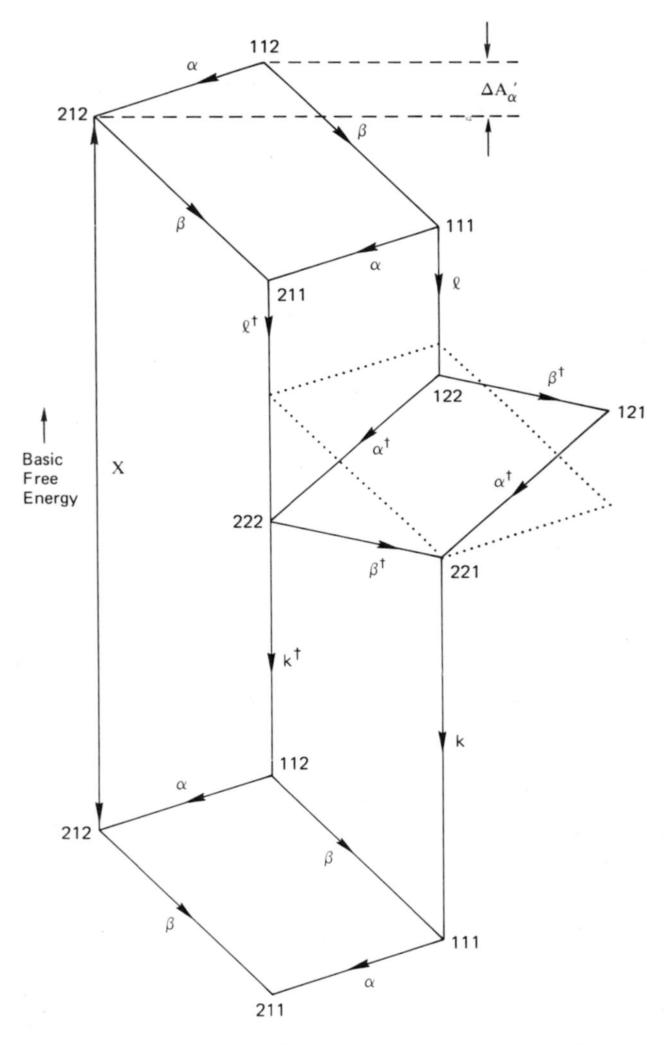

FIG. 7.14 Basic free energy levels (hypothetical) corresponding to Fig. 7.13. See text for details.

There are two independent connections between rate constants or free energies. These follow from

$$e^{X/kT} = \Pi_{\kappa+}/\Pi_{\kappa-} \qquad (\kappa = a, b, \ldots, f) \tag{7.27}$$

or, more simply, from Fig. 7.14 (compare the two possible paths between 111 and 222 and between 222 and 111):

$$\alpha_{12} l_{12}^{\dagger} / \alpha_{21} l_{21}^{\dagger} = l_{12} \alpha_{12}^{\dagger} / l_{21} \alpha_{21}^{\dagger}$$
$$k_{21}^{\dagger} \beta_{21} / k_{12}^{\dagger} \beta_{12} = \beta_{21}^{\dagger} k_{21} / \beta_{12}^{\dagger} k_{12} \tag{7.28}$$

$$\Delta A'_{\alpha} + \Delta A'^{\dagger}_{l} = \Delta A'_{l} + \Delta A'^{\dagger}_{\alpha}$$
$$\Delta A'^{\dagger}_{k} + \Delta A'_{\beta} = \Delta A'^{\dagger}_{\beta} + \Delta A'_{k}. \tag{7.29}$$

For interpretation in terms of interaction free energies, see the end of Section 7.3.

Because this is effectively a single-cycle system with a single flux J^{∞} and a single net force X, free energy transduction is uncomplicated—as discussed in the paragraph following Eq. 7.7. Also, it is for the same reason that a single set of free energy levels (Fig. 7.14) suffices for this system, rather than each cycle requiring its own set.

Extended Scheme with No Waiting

This is essentially an appendix, which will be found useful in Section 7.4. We consider an interesting kinetic question. Suppose we have $m + 1$ two-state enzymes in a linear array, with no back reactions, with no waiting time for the two single reactions at the ends of the array (i.e., these reactions are assumed to be relatively fast), and with the *same* rate constant k for all m interlocking (coupled) reactions. What is J^{∞}/N for arbitrary m? Actually, the symmetry is different for m even and odd, so that there are really *two* general problems.

Results already given above take care of the cases $m = 1, 2, 3$, and 4. For $m = 1$, we obviously have $J^{\infty}/N = k$. This also follows from Eq. 7.10 on putting $\alpha, \alpha^{\dagger}, \beta, \beta^{\dagger} \to \infty$. For $m = 2$, we have, in effect, a two-state one-way cycle with both rate constants equal to k. Equation 7.23 applies with $l = k$. Hence, $J^{\infty}/N = k/2$. This is also a consequence of Eq. 7.22 ($\alpha, \beta \to \infty, l = k$) and of Eq. 7.10 ($\alpha = \alpha^{\dagger} \to \infty$, $\beta = \beta^{\dagger} = k$). For $m = 3$, we put $\alpha = \alpha^{\dagger} = \beta = \beta^{\dagger} = k$ in Eq. 7.10 and find $J^{\infty}/N = 2k/5$. The same result is obtained from Eq. 7.22 if we let $\alpha \to \infty$ and take $\beta = l = k$. Finally, for $m = 4$, we set $\alpha = \beta = l = k$ in Eq. 7.22 and get $J^{\infty}/N = 5k/14$.

We have gone (1) four steps ($m = 5$ to 8) beyond the above results by adding up to four more enzymes to Fig. 7.10a and taking all rate constants equal to k. There are $2^4 = 16$ states for $m = 5$ and $2^7 = 128$ states for $m = 8$. Tedious but simple calculations (1) (using essentially the steady-state algebraic equations, but not cycles) lead to $J^{\infty}/N = k/3, 7k/22, 4k/13$, and $3k/10$, respectively.

The general formula for J^{∞}/Nk for any m, odd or even, is apparently $(m + 1)/2(2m - 1)$, though we have not proved this. As $m \to \infty$, $J^{\infty}/Nk \to \frac{1}{4}$.

This result is as expected for a random assignment of each subunit between its two states.

For a *closed ring* of m interlocking reactions (this model was suggested by Dr. Igor Plesner), the corresponding result (1) for J^∞/Nk is $m/4(m-1)$, where m is even and also the numbers of subunits in states 1 and 2 are even. This result is easy to prove (and generalize). But we omit details (1) here because this case probably has no biological significance.

7.3 Example: Two Interacting Enzymes

The beginning of Section 7.1 should be read as an introduction to this section. Figure 7.3 was introduced there and it is this case that we now investigate.

We consider two enzymes or macromolecules α and β, in contact, that influence each other's kinetics, though there are no interlocking reactions as in our definition of an "enzyme complex": the interactions are of type (b) above. The thermodynamic forces are X_α and X_β and the fluxes are J_α^∞ and J_β^∞. Both cycles have been reduced (Appendix 1) to two-state cycles (e.g., Fig. 7.1 → Fig. 7.3). In deriving kinetic properties, we again assume one-way reactions for simplicity. Back reactions will be included in the next subsection in order to treat the basic free energy levels.

The diagram is shown in Fig. 7.15a. Note that the transitions with an asterisk ($2 \to 1$) replace the k transition ($22 \to 11$) in Fig. 7.7. (In this section, an asterisk does *not* mean a second-order rate constant.) That is, each enzyme can now *separately* undergo the transition $2 \to 1$, thus completing a cycle. The asterisk transitions are *not* inverses of the other transitions shown (rather, β_{21} is inverse to $\beta \equiv \beta_{12}$, etc.).

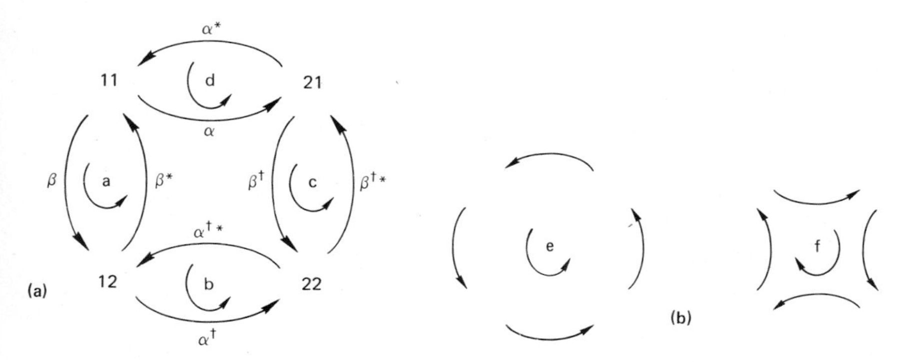

FIG. 7.15 (a) Diagram (with rate constant notation—back reactions omitted) based on Fig. 7.3. (b) Cycles e and f for this system (cycles a, b, c, d are indicated in Fig. 7.15a).

This system has six cycles with nonzero force (if back reactions are included, there are 16 such cycles). These are labeled *a*, *b*, ..., *f* in Figs. 7.15a and 7.15b. The force in cycles *a* and *c* is X_β, in *b* and *d* it is X_α, and in *e* and *f* it is $X_\alpha + X_\beta$. As we saw in Chapter 3, the mere fact that there is at least one cycle that contains both forces guarantees that there will, in general, be coupling between the fluxes and, therefore, free energy transduction. But this coupling will not be *complete*, as it is in the linear multienzyme complexes we have been considering (Fig. 7.9). The coupling will, of course, disappear if the two enzymes do not interact with each other at all, that is, if the dagger symbol can be dropped from Fig. 7.15a ($\beta = \beta^\dagger$, $\alpha^* = \alpha^{\dagger *}$, etc.). If back reactions are included, four of the 16 nonzero-force cycles contain both forces.

Each of cycles *a*–*d* has three flux diagrams, while cycles *e* and *f* have only one each. Hence,

$$\begin{aligned}
J_a &= N\beta\beta^*(\alpha^*\beta^{\dagger *} + \beta^\dagger\alpha^{\dagger *} + \alpha^*\alpha^{\dagger *})/\Sigma, \\
J_b &= N\alpha^\dagger\alpha^{\dagger *}(\alpha^*\beta + \alpha\beta^\dagger + \beta\beta^\dagger)/\Sigma \\
J_c &= N\beta^\dagger\beta^{\dagger *}(\alpha\beta^* + \alpha^\dagger\beta + \alpha\alpha^\dagger)/\Sigma, \\
J_d &= N\alpha\alpha^*(\alpha^{\dagger *}\beta^* + \alpha^\dagger\beta^{\dagger *} + \beta^*\beta^{\dagger *})/\Sigma \\
J_e &= N\beta\alpha^\dagger\beta^{\dagger *}\alpha^*/\Sigma, \qquad J_f = N\alpha\beta^\dagger\alpha^{\dagger *}\beta^*/\Sigma,
\end{aligned} \tag{7.30}$$

where Σ is to be determined below.

Once again the steady-state probabilities follow quite simply from the flux diagrams (Section 2.4). Figure 7.16 shows the composition of each transition flux in terms of cycles. Thus, for example,

$$\begin{aligned}
\beta p_{11}^\infty &= (J_a + J_e)/N, \qquad & \alpha^\dagger p_{12}^\infty &= (J_b + J_e)/N \\
\beta^{\dagger *} p_{22}^\infty &= (J_c + J_e)/N, \qquad & \alpha^* p_{21}^\infty &= (J_d + J_e)/N.
\end{aligned} \tag{7.31}$$

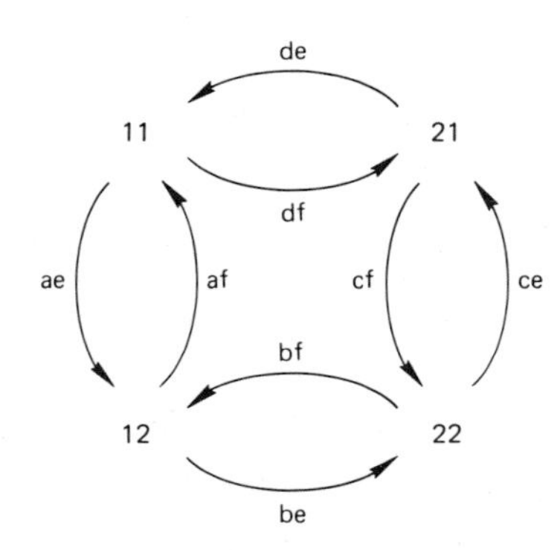

FIG. 7.16 Listing of cycles that contribute to each transition flux in Fig. 7.15a.

Using Eqs. 7.30, we easily obtain the p_{ij}^{∞} with Σ still unspecified. Σ is then found by normalization:

$$\begin{aligned}\Sigma = {} & \alpha^{\dagger *}\beta^{*}(\alpha^{*} + \beta^{\dagger}) + \alpha^{*}\beta^{\dagger *}(\alpha^{\dagger} + \beta^{*}) \\ & + \alpha^{*}\beta(\alpha^{\dagger *} + \beta^{\dagger *}) + \alpha^{\dagger *}\beta^{\dagger}(\alpha + \beta) \\ & + \alpha\beta^{*}(\alpha^{\dagger *} + \beta^{\dagger *}) + \alpha^{\dagger}\beta^{\dagger *}(\alpha + \beta) \\ & + \alpha\beta^{\dagger}(\alpha^{\dagger} + \beta^{*}) + \alpha^{\dagger}\beta(\alpha^{*} + \beta^{\dagger}). \end{aligned} \tag{7.32}$$

Actually, the terms are arranged here in the order p_{11}^{∞}, p_{12}^{∞}, p_{21}^{∞}, p_{22}^{∞}, so Eq. 7.32 also gives the p_{ij}^{∞}.

The separate α and β steady-state fluxes are, from Fig. 7.16,

$$J_{\alpha}^{\infty} = J_b + J_d + J_e + J_f, \qquad J_{\beta}^{\infty} = J_a + J_c + J_e + J_f. \tag{7.33}$$

These can be written out explicitly, if desired, using Eqs. 7.30 and 7.32.

In the special case that there are no $\alpha\beta$ interactions (omit all daggers), these results simplify to

$$\Sigma = (\alpha + \alpha^{*})(\beta + \beta^{*})(\alpha + \alpha^{*} + \beta + \beta^{*}) \tag{7.34}$$

$$p_{11}^{\infty} = \frac{\alpha^{*}}{\alpha + \alpha^{*}} \cdot \frac{\beta^{*}}{\beta + \beta^{*}}, \qquad p_{12}^{\infty} = \frac{\alpha^{*}}{\alpha + \alpha^{*}} \cdot \frac{\beta}{\beta + \beta^{*}}, \tag{7.35}$$

etc. Also,

$$J_{\alpha}^{\infty} = N\alpha\alpha^{*}/(\alpha + \alpha^{*}), \tag{7.36}$$

etc. This is just what we expect for independent subunits.

As another special case (an extreme form of interaction), suppose enzyme β can operate only if enzyme α is in state 1. That is, $\beta^{\dagger} = 0$ and $\beta^{\dagger *} = 0$ in Fig. 7.15a. The diagram now has only the three cycles a, b, and d. No thermodynamic coupling or free energy transduction is possible between α and β because no cycle contains both X_{α} and X_{β}. We have in this case

$$\Sigma = \alpha^{\dagger *}\beta^{*}\alpha^{*} + \alpha^{*}\beta\alpha^{\dagger *} + \alpha\beta^{*}\alpha^{\dagger *} + \alpha^{\dagger}\beta\alpha^{*}, \tag{7.37}$$

where the terms are again listed in the order 11, 12, 21, 22 (to give the p_{ij}^{∞}). The fluxes are

$$J_{\alpha}^{\infty} = J_b + J_d = N\alpha^{\dagger *}\alpha^{*}(\alpha^{\dagger}\beta + \alpha\beta^{*})/\Sigma, \qquad J_{\beta}^{\infty} = J_a = N\beta\beta^{*}\alpha^{*}\alpha^{\dagger *}/\Sigma. \tag{7.38}$$

Although the enzymes influence each other, as seen in these equations, one cannot "drive" the other. For example (with back reactions included), if $X_{\alpha} = 0$, a force $X_{\beta} \neq 0$ cannot induce a flux $J_{\alpha}^{\infty} \neq 0$.

In a still more extreme case, suppose *neither* enzyme operates unless the other is in state 1. Thus we also put $\alpha^\dagger = 0$ and $\alpha^{\dagger *} = 0$. The diagram has cycles a and d only. The steady-state properties are

$$\begin{gathered} p_{11}^\infty = \alpha^*\beta^*/\Sigma, \qquad p_{12}^\infty = \alpha^*\beta/\Sigma, \qquad p_{21}^\infty = \alpha\beta^*/\Sigma \\ J_\alpha^\infty = J_d = N\alpha\alpha^*\beta^*/\Sigma, \qquad J_\beta^\infty = J_a = N\alpha^*\beta^*\beta/\Sigma \\ \Sigma = \alpha^*\beta^* + \alpha^*\beta + \alpha\beta^*. \end{gathered} \tag{7.39}$$

In this model only one enzyme can work (cycle) at a time. Hence both J_α^∞ and J_β^∞ are smaller than the "independent enzyme" fluxes given by Eq. 7.36, etc.

Free Energy Levels

To discuss the thermodynamics of the model represented by Fig. 7.15a, we now include all back reactions: β becomes β_{12} and we add β_{21}; etc. The cycles a, b, c, d then lead to

$$\begin{gathered} \beta_{12}\beta_{21}^*/\beta_{21}\beta_{12}^* = \Pi_{a+}/\Pi_{a-} = \Pi_{c+}/\Pi_{c-} = e^{X_\beta/kT} \\ \Pi_{b+}/\Pi_{b-} = \Pi_{d+}/\Pi_{d-} = e^{X_\alpha/kT}. \end{gathered} \tag{7.40}$$

These provide two required relations between rate constants. The corresponding basic free energy relations are obvious from Fig. 7.17 and need not be repeated in the text. There is, however, one additional free energy relation that is not obvious: the two double-headed arrows in Fig. 7.17 must have the same length. This can be seen, for example, from cycle e and Fig. 7.18.

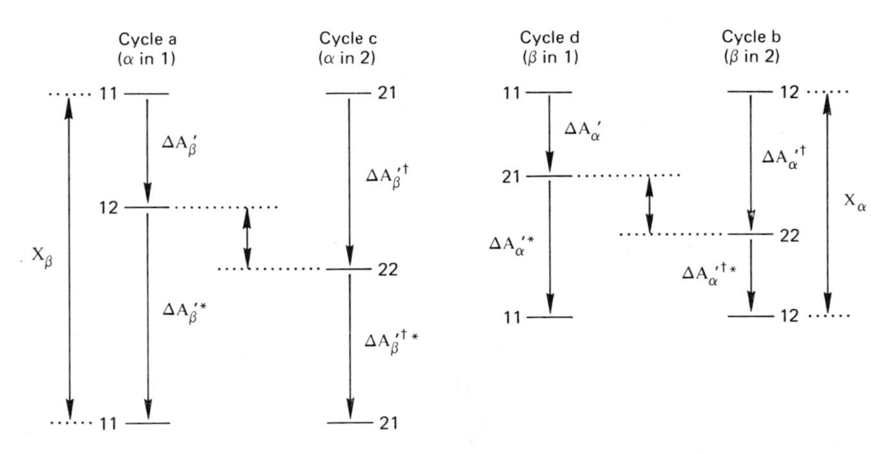

FIG. 7.17 Basic free energy levels in cycles a, b, c, d of Fig. 7.15a (with back reactions included). See text for details.

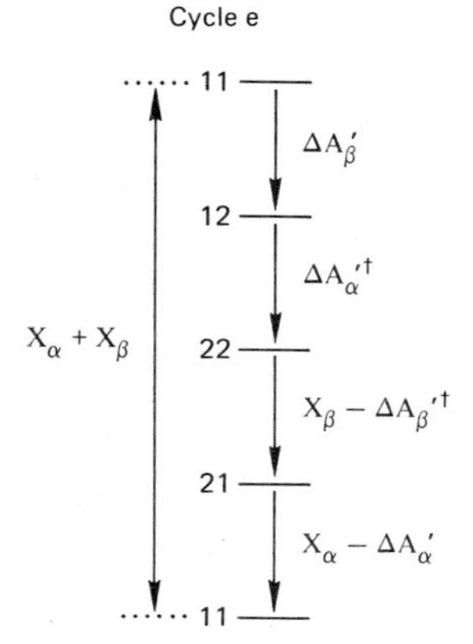

FIG. 7.18 Basic free energy levels for cycle *e* of Fig. 7.15b.

The free energy differences indicated in Fig. 7.18 follow from Fig. 7.17. On adding these differences, we find

$$\Delta A_\beta'^\dagger - \Delta A_\beta' = \Delta A_\alpha'^\dagger - \Delta A_\alpha', \tag{7.41}$$

which is one way of stating that the double-headed arrows in Fig. 7.17 are equal in length. Another, not independent, way is

$$\Delta A_\beta'^* - \Delta A_\beta'^{\dagger *} = \Delta A_\alpha'^* - \Delta A_\alpha'^{\dagger *}. \tag{7.42}$$

To understand the physical significance of Eq. 7.41, let us introduce—for the first time in this chapter—interaction free energies explicitly. We denote the interaction free energies between enzymes α and β, for all possible states ij, by w_{11}, w_{12}, w_{21}, and w_{22}. The free energies w_{ij} belong to $\alpha\beta$ *pairs* and cannot be broken down into α and β parts. If the enzymes α and β are physically separated, all $w_{ij} \to 0$. Further, we designate the basic free energy changes for *separated* enzymes α and β as in Fig. 7.19. Note that X_α and X_β would not be affected by the separation. Then, for example,

$$\begin{aligned}
&(11 \to 12) \qquad \Delta A_\beta' = \Delta A_\beta^0 + w_{11} - w_{12} \\
&(21 \to 22) \qquad \Delta A_\beta'^\dagger = \Delta A_\beta^0 + w_{21} - w_{22} \\
&(11 \to 21) \qquad \Delta A_\alpha' = \Delta A_\alpha^0 + w_{11} - w_{21} \\
&(12 \to 22) \qquad \Delta A_\alpha'^\dagger = \Delta A_\alpha^0 + w_{12} - w_{22}.
\end{aligned} \tag{7.43}$$

It follows from these relations that (Eq. 7.41)

$$\begin{aligned}
\Delta A_\beta'^\dagger - \Delta A_\beta' &= \Delta A_\alpha'^\dagger - \Delta A_\alpha' \\
&= w_{12} + w_{21} - (w_{11} + w_{22}).
\end{aligned} \tag{7.44}$$

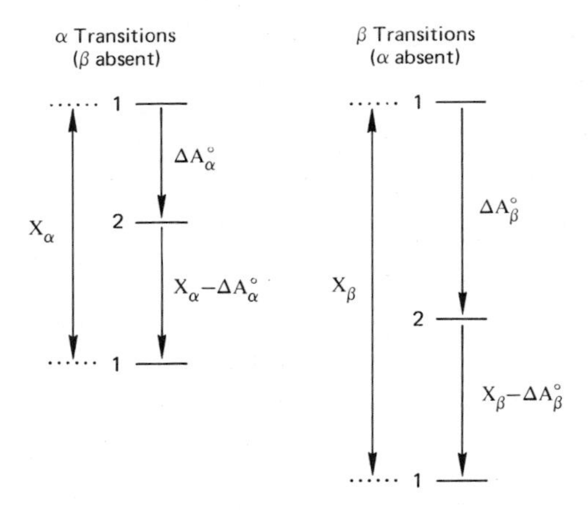

FIG. 7.19 Basic free energy levels for the model in Fig. 7.15 if the two enzymes are *separated* (noninteracting). See text for details.

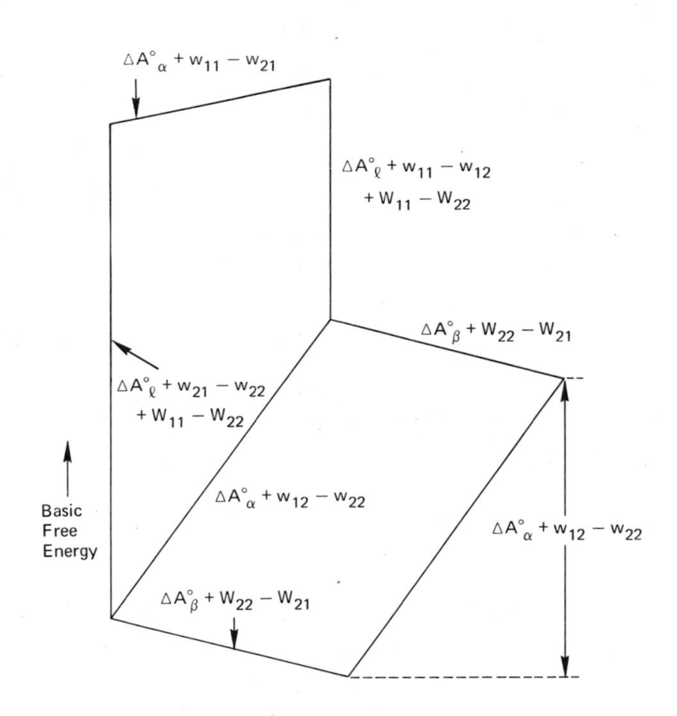

FIG. 7.20 Part of Fig. 7.14 with basic free energy changes (vertical direction) broken down into interaction contributions. See text.

This is the change in interaction free energy in the exchange of partners

$$1_\alpha 2_\beta + 2_\alpha 1_\beta \rightarrow 1_\alpha 1_\beta + 2_\alpha 2_\beta . \tag{7.45}$$

If we introduce the w_{ij} into Section 7.1 in the same way, each side of Eq. 7.3 becomes equal to $\Delta A_\alpha^0 + \Delta A_\beta^0 + w_{11} - w_{22}$.

For the model studied in Section 7.2 (see Figs. 7.10a and 7.14), we need two sets of interaction free energies: w_{ij} for $\alpha\gamma$ interactions and W_{ij} for $\gamma\beta$ interactions. Although the situation is more complicated, no really new features emerge. As an illustration, part of Fig. 7.14 is repeated in Fig. 7.20, with basic free energy changes written as in Eqs. 7.43. The reader may wish to complete this figure. The hypothetical noninteraction basic free energy changes must satisfy

$$\Delta A_\alpha^0 + \Delta A_\beta^0 + \Delta A_l^0 + \Delta A_k^0 = X. \tag{7.46}$$

7.4 Oxidative Phosphorylation

In this section we give a *simplified* sketch of the free energy transduction that occurs in oxidative phosphorylation, from the point of view of the present book. We adopt, as a basis for the discussion, the general principles of the chemiosmotic hypothesis of Mitchell (14), as summarized recently by Racker (15). However, for our limited purposes, we need not be very explicit about details of the mechanism, stoichiometry, etc. The Mitchell hypothesis is, of course, not the only hypothesis under active discussion in this field (13).

There are two independent multienzyme complexes to consider (15): the respiratory chain enzymes and the proton driven reverse-ATPase. There are thousands of units (13) of both types in the inner membrane of a mitochondrion. Though independent, the two types of units operate simultaneously, along with other membrane activity of various sorts, producing steady-state concentrations, on the two sides of the inner membrane, of all substrates, products, ligands, etc., involved in the workings of these units, as well as a steady-state membrane potential.

The respiratory complex is, or is not very different from, a linear complex of the type discussed in Section 7.2. We shall treat it here as linear. The single *net* thermodynamic force X driving the "collapsed-single-cycle" flux (Fig. 7.9) is comprised of two coupled components: the chemical reaction (a sum of two half-reactions)

$$\mathrm{NADH} + \tfrac{1}{2}\mathrm{O}_2 + \mathrm{H}^+ \rightarrow \mathrm{NAD}^+ + \mathrm{H}_2\mathrm{O} \qquad (X_1 > 0) \tag{7.47}$$

and proton transport

$$m_1\mathrm{H}^+ \text{ (in; matrix)} \rightarrow m_1\mathrm{H}^+ \text{ (out)} \qquad (X_2 < 0), \tag{7.48}$$

where $X_1 > -X_2$. That is, the *net* force $X = X_1 + X_2$ is positive for the reactions as written (in normal operation). X_1 is the Gibbs free energy change at *actual* (not standard) steady-state concentrations, as in Eqs. 3.30 and 3.38. Also, $X_2 = m_1(\mu_{H^+}^{in} - \mu_{H^+}^{out})$, where ( ) is the electrochemical potential difference for protons at actual steady-state concentrations and membrane potential. The net result of the cyclic action of the respiratory complex is for reaction 7.47 to drive protons uphill in → out (Eq. 7.48). The value of m_1 is somewhat uncertain; it is probably between 3×3 and 3×4 (16).

The mechanism of action of the reverse-ATPase is not well established yet (15). It is probably a multienzyme complex in the sense of the present chapter but it may or may not have a linear sequence (as in Section 7.2). In the independent but simultaneous activity of this complex, proton transport

$$m_2\,H^+\ (\text{out}) \rightarrow m_2\,H^+\ (\text{in}) \qquad (X_3 > 0) \tag{7.49}$$

drives the synthesis of ATP,

$$3\text{ADP} + 3\text{P}_i \rightarrow 3\text{ATP} \qquad (X_4 < 0). \tag{7.50}$$

Although presumably $m_1 \cong m_2$, these numbers need not be exactly equal because of other concurrent steady-state inner membrane activity that could involve protons. A value for m_2 of approximately 3×3, or more, seems necessary (17) in order to achieve a positive net force $X_3 + X_4 > 0$, as required. Of course the (proton) forces X_2 and X_3 are closely related since both complexes operate from the same steady-state proton pools: $-X_2/m_1 = X_3/m_2$.

In effect, the proton gradient acts here (14, 15) as a common intermediate (see the end of Section 3.2), coupling Eq. 7.47 to Eq. 7.50. The overall process is analogous, in reverse, to the establishment of a Na^+ gradient by hydrolysis of ATP (Na,K-ATPase) which is used to drive, in a separate enzymatic operation, an amino acid, say, uphill (out → in) across the same membrane.

Respiratory Chain

The remainder of the section is concerned only with the respiratory complex of enzymes, simplified and linear. The reactions are indicated very schematically (for two electrons) in Fig. 7.21, where O stands for oxidized form and R for reduced form of the enzymes I, II, ... All reactions are "interlocking" except the two end reactions. The m_1 protons in Eq. 7.48 have to enter and leave the scheme, but this feature is not included explicitly in the figure. There are M enzymes in the complex; each is shown as having two possible states though more would actually be necessary in some cases (e.g., to accommodate the protons). In the actual respiratory chain, M is of order 15 or more [Chance (15)].

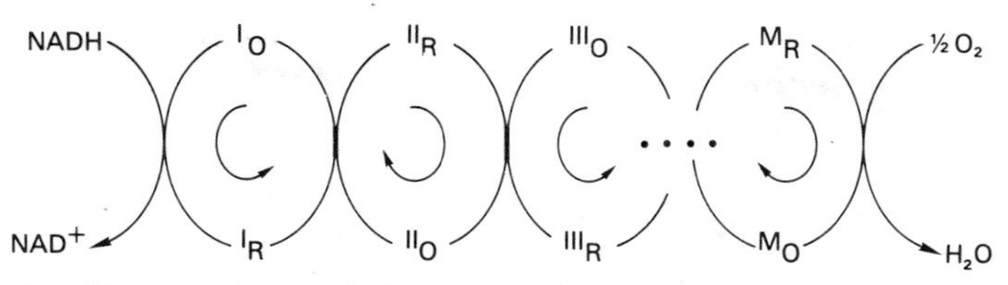

FIG. 7.21 Linear respiratory chain complex of M two-state enzymes. O means oxidized; R, reduced. The heavy bars indicate "interlocking" reactions.

The state of a unit (complex) is specified only when the state of each subunit (enzyme) is specified (O or R). The complex has a total of 2^M possible states.*

To illustrate a basic free energy change, suppose the transition pair (left end of Fig. 7.21)

$$\mathrm{OOO}\cdots \underset{\alpha'}{\overset{\alpha}{\rightleftharpoons}} \mathrm{ROO}\cdots \tag{7.51}$$

corresponds to the (balanced) reaction

$$\mathrm{NADH} + \mathrm{H}^+ + \mathrm{I} \to \mathrm{NAD}^+ + \mathrm{IH}_2, \tag{7.52}$$

where $\mathrm{I} \equiv \mathrm{I_O}$ and $\mathrm{IH_2} \equiv \mathrm{I_R}$ in Fig. 7.21. Some transient intermediate states are obviously omitted in Eq. 7.52. Then, for this process (see Eq. 4.19 and Table 4.1),

$$\begin{aligned} \Delta A' &= (A_{\mathrm{I}} + \mu_{\mathrm{NADH}} + \mu_{\mathrm{H}^+}) - (A_{\mathrm{IH}_2} + \mu_{\mathrm{NAD}^+}) \\ &= kT \ln(\alpha/\alpha'). \end{aligned} \tag{7.53}$$

Both α and α' are pseudo-first-order rate constants. The μ's refer to actual (not standard) concentrations. Obviously, oxidation–reduction reactions do not require any modification of the previous formalism (because half-reactions always occur in pairs).

In several places in the respiratory chain, the interlocking transition consists of a transfer of a pair of electrons only (13):

$$\begin{gathered} \cdots\mathrm{OROO}\cdots \underset{k'}{\overset{k}{\rightleftharpoons}} \cdots\mathrm{OORO}\cdots \\ \Delta A' = A_{\mathrm{OROO}} - A_{\mathrm{OORO}} = k_{\mathrm{B}} T \ln(k/k'), \end{gathered} \tag{7.54}$$

* This type of kinetic scheme for the respiratory chain was introduced by W. F. Holmes in his Ph.D. thesis (University of Pennsylvania, 1960). Holmes studied, especially, transients and the "mass-action" (see p. 190) approximation for the cases $M = 2$ and $M = 3$.

where k_B = Boltzmann constant. As no ligands, substrates, etc., are involved in such a reaction, from the point of view of the complex these are just isomeric transitions (Table 4.1) with first-order rate constants.

As a first explicit and simple example, let us adopt Fig. 7.10 to the present context, obtaining Fig. 7.22. That is, suppose that the "respiratory chain" has three enzymes I, II, and III ($M = 3$), that the two end reactions (Fig. 7.21) have (first-order) rate constants α ($O \rightarrow R$) and β ($R \rightarrow O$), and that there are two interlocking reactions with rate constants k and l ($RO \rightarrow OR$ for both). To ease the algebra, we are assuming, it will be recalled

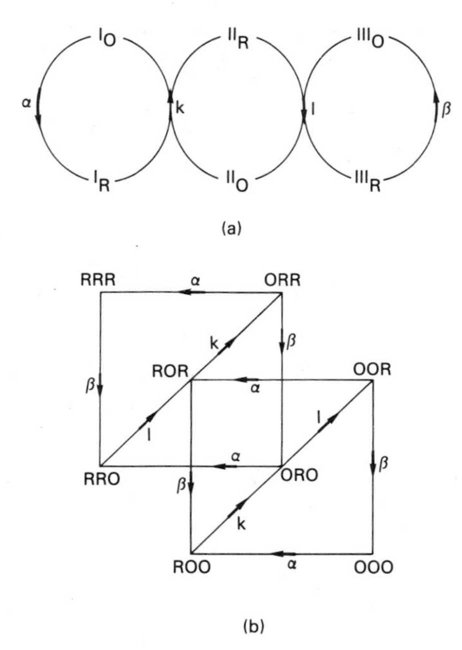

FIG. 7.22 (a) Complex of three two-state enzymes, with one-way reactions. (b) Corresponding diagram.

(Fig. 7.10), that back reactions can be neglected and that there are no interactions of type (b) (Section 7.2) between neighboring enzymes of the chain. Equations 7.20–7.22 give the steady-state properties of this system (the notation has to be translated, as in Fig. 7.10 → 7.22).

Note the progression, in Fig. 7.22b, from all-reduced (RRR, upper left) to all-oxidized (OOO, lower right).

In order to get a qualitative feel for the nature of this steady state, let us simplify even further by taking all rate constants equal to k ($\alpha = \beta = k = l$), as at the end of Section 7.2 (1). Then Eqs. 7.20 give the results in Fig. 7.23.

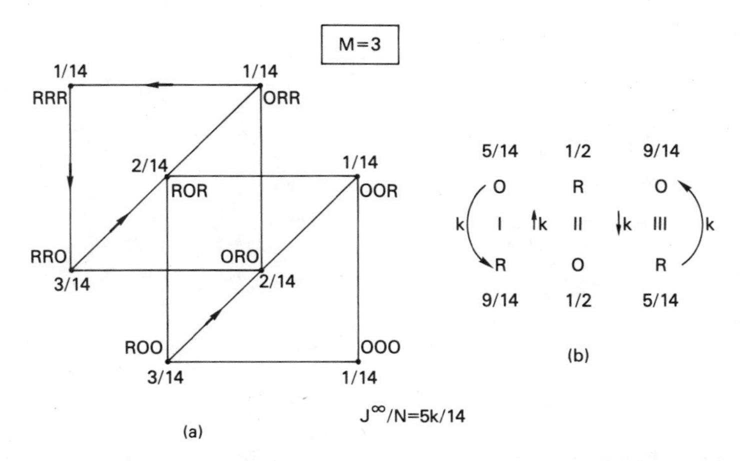

FIG. 7.23 (a) Steady-state probabilities of the states of an $M = 3$ complex, when all rate constants (one-way) are equal to k. Arrows show the directions of horizontal, vertical, and diagonal transitions. (b) Steady-state probabilities of the states of individual subunits, found by combining probabilities in (a).

The steady-state probabilities of the eight states RRR, etc., are given in Fig. 7.23a. These *unit* (i.e., enzyme complex) probabilities can be combined to obtain the separate *subunit* (i.e., enzyme) probabilities in Fig. 7.23b. For example, for enzyme I,

$$P_{\mathrm{O}}^{\infty} = \tfrac{1}{14} + \tfrac{2}{14} + \tfrac{1}{14} + \tfrac{1}{14} = \tfrac{5}{14}, \qquad P_{\mathrm{R}}^{\infty} = \tfrac{1}{14} + \tfrac{3}{14} + \tfrac{2}{14} + \tfrac{3}{14} = \tfrac{9}{14}.$$

The probability P_{R}^{∞} of the reduced state, for I, II, and III, respectively, decreases: $\frac{9}{14}, \frac{1}{2}, \frac{5}{14}$. The average value is $\frac{1}{2}$. This is a simple illustration of the progressively decreasing degree of reduction in the respiratory chain (in the direction $\mathrm{NADH} \rightarrow \mathrm{O_2}$), in so-called state 4 respiration (18). Note that the flux $J^{\infty}/N = 5k/14$ follows directly from P_{O}^{∞} for subunit I (or from P_{R}^{∞} for III).

Kinetically, this steady-state subunit (enzyme) probability gradient (with all rate constants equal) is due to the bias introduced by the two end reactions. That is, it is an end effect. The reaction at the left end favors state R for enzyme I while the reaction at the right end favors state O for enzyme III. This asymmetry or bias disappears in the "closed ring" model (M even) mentioned at the end of Section 7.2 (1).

The probability "gradient" (lower left to upper right) in Fig. 7.23a is of course also due to the same end reaction bias.

Figures 7.24 (based on Eqs. 7.12), 7.25, and 7.26 (based on the steady-state algebraic equations—see the end of Section 7.2), extend the results in Fig. 7.23 to $M = 2$, 4, and 5, respectively (1). The states in Fig. 7.24 are

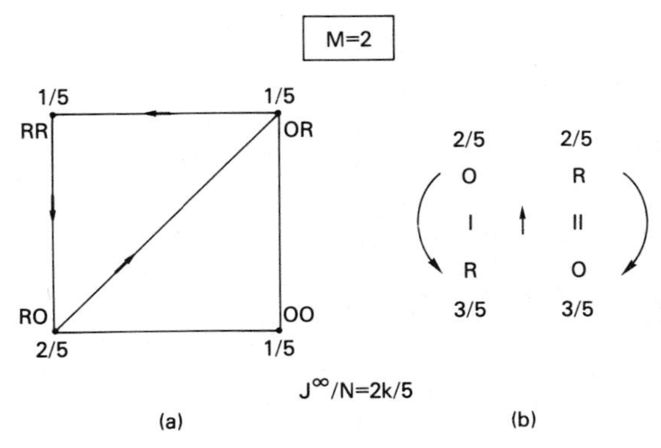

FIG. 7.24 Same as Fig. 7.23, but for $M = 2$.

arranged differently than in Figs. 7.4b and 7.7, in order to be consistent with Figs. 7.23, 7.25, and 7.26. In all of these figures, the number of states in the diagram is 2^M and the number of squares is 2^{M-2}. The same gradients in unit and subunit probabilities as in Fig. 7.23 can be seen. Because of the connection between J^∞/Nk and P_O^∞ for enzyme I, already mentioned, and since (Section 7.2; $M + 1 \equiv m$)

$$J^\infty/Nk = (M + 2)/2(2M + 1) \to \tfrac{1}{4} \qquad (M \to \infty), \tag{7.55}$$

we have

$$\begin{aligned} P_R^\infty(\text{enzyme I}) = 1 - (J^\infty/Nk) &= 3M/2(2M + 1) \\ &\to \tfrac{3}{4} \qquad (M \to \infty). \end{aligned} \tag{7.56}$$

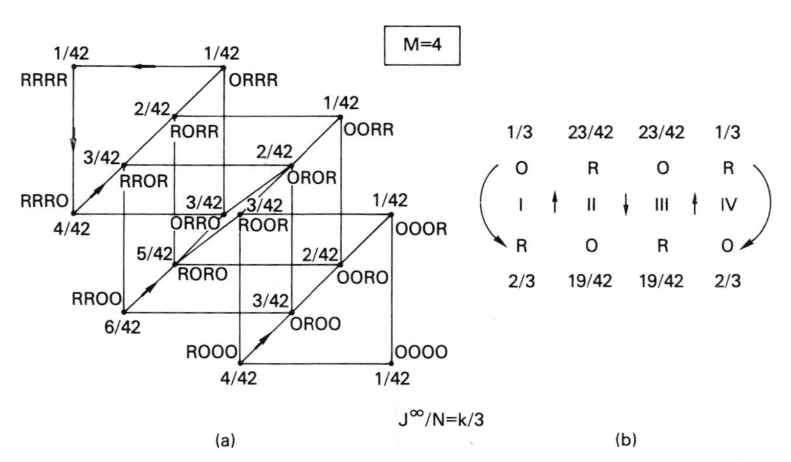

FIG. 7.25 Same as Fig. 7.23, but for $M = 4$.

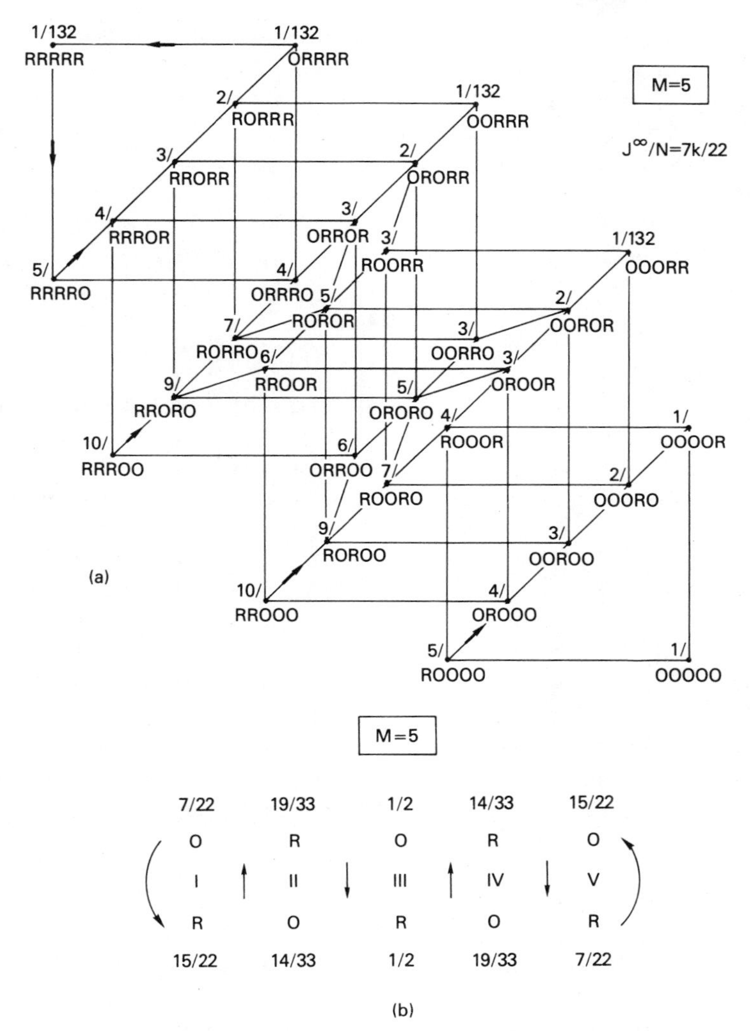

FIG. 7.26 Same as Fig. 7.23, but for $M = 5$. The denominator 132 is omitted from most probabilities in (a).

Equation 7.55 also gives P_{R}^{∞} for enzyme M (the last in the chain). Of course these *quantitative* results hold only for a very special choice of rate constants.

The solution of the steady-state algebraic equations can be verified in Figs. 7.23a–7.26a. As an example, consider state ROORO in Fig. 7.26a. The sum of probabilities entering this state is $(9/132) + (2/132) + (3/132)$ while the sum leaving the state is $(7/132) + (7/132)$. The two sums must be equal at steady state (because all rate constants are equal in this example).

Incidentally, if in this model we allow only the single sequence of transitions and states in the lower right-hand cycle of each of Figs. 7.23a–7.26a (i.e., OOO $\cdots \rightarrow$ ROO $\cdots \rightarrow$ ORO $\cdots$, etc.), then we have a single-cycle, one-way diagram with all rate constants equal to k. All state probabilities, at steady-state, are then equal to $1/(M+1)$ and the flux is $J^{\infty}/Nk = 1/(M+1)$. Note that this flux goes to zero as $M \rightarrow \infty$, while the correct asymptotic limit is $\frac{1}{4}$ (in Eq. 7.55). That is, in the actual system (complete diagram), there is no particular disadvantage, for the flux, if the chain of enzymes is very long; but this *would* be disadvantageous in the hypothetical single-cycle system. In the complete diagram, as M increases, multiple paths cancel the effect of length of chain. For $M = 10$, the two fluxes are 2/7 (complete diagram) and 1/11 (single cycle), with a ratio of 22/7.

The relative basic free energy levels of the states of a complex can be discussed only if back reactions are included. To maintain maximum simplicity, suppose the forward rate constant for every transition is still k, as above, while the backward rate constant is always k'. (Actually, all we require here is that the rate constant *ratio* for every transition have the same value, k/k'.) To be explicit, we take $k > k'$ (the case of primary interest). The basic free energy drop for every transition is then $\Delta A' = k_B T \ln(k/k')$. Furthermore, the net thermodynamic force in the respiratory chain is $X = X_1 + X_2 = (M+1)\,\Delta A'$, since there are $M+1$ transitions and states in the simplest cycles of the diagram. For $M = 3$, Fig. 7.14 is symmetrized to appear as in Fig. 7.27. Figure 7.28 shows the corresponding set of levels for $M = 4$. As in Fig. 7.14, vertical transitions in Figs. 7.27 and 7.28 are "interlocking" (k, l, m and reverse) while diagonal transitions represent the end reactions (α, β and reverse). When M is even, all-R states are at the same free energy levels as all-O states. But when M is odd, all-R and all-O are staggered, with a separation in free energy of $\pm X/2$.

Any single system (complex) does a biased "walk" on the basic free energy levels of Fig. 7.28 (if $M = 4$), the general trend of the walk being downhill ($k > k'$). The levels extend indefinitely, above and below those shown.

Free energy transduction is accomplished by these enzyme complexes, with efficiency $\eta = -X_2/X_1$ (Eqs. 7.47 and 7.48). This transduction cannot be localized in particular transitions of, say, Fig. 7.28 (even if we knew all the details of the mechanism) but, as usual, is a property of the *entire* system of reactions and free energy levels.

Flux with Back Reactions

When M is small, it is not difficult to deduce the steady-state probabilities and flux when all forward rate constants are k and all backward rate

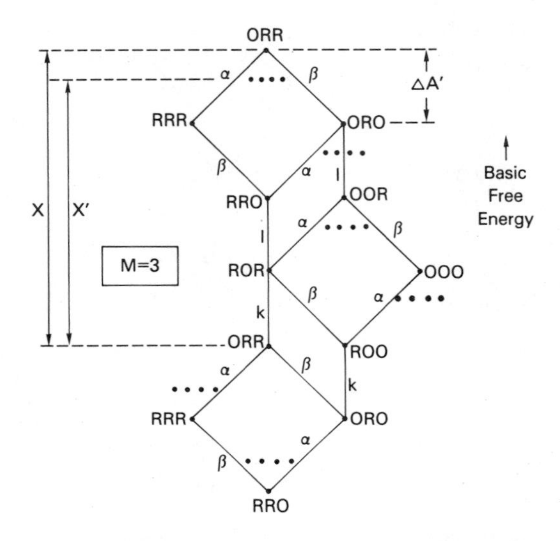

FIG. 7.27 Basic free energy levels in $M = 3$ case when all the transitions in the directions shown in Fig. 7.22b have the same basic free energy drop $\Delta A' = k_B T \ln(k/k')$, where k_B is the Boltzmann constant. See text. This is a special case of Fig. 7.14. Dotted levels and X' arise as a result, say, of lowering the concentration of NADH.

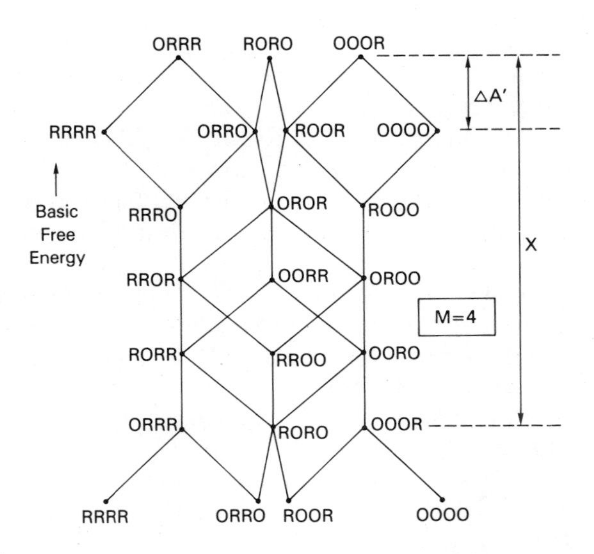

FIG. 7.28 Same as in Fig. 7.27, but for $M = 4$.

constants are k' (1). For $M = 1$, we find

$$J^{\infty}/N = (k - k')/2. \tag{7.57}$$

For $M = 2$,

$$J^{\infty}/N = 2(k^3 - k'^3)/(5k^2 + 6kk' + 5k'^2). \tag{7.58}$$

For $M = 3$,

$$\frac{J^{\infty}}{N} = \frac{5(k^4 - k'^4) + 6kk'(k^2 - k'^2)}{(5k + 3k')(k^2 + 2kk' + 3k'^2) + (3k + 5k')(3k^2 + 2kk' + k'^2)}. \tag{7.59}$$

Since

$$X = (M + 1)\,\Delta A' = (M + 1)k_B T \ln(k/k'), \tag{7.60}$$

we obtain, *near equilibrium* ($k = k'$ at equilibrium),

$$\begin{aligned} M = 1&: \quad J^{\infty}/N \to kX/4k_B T \\ M = 2&: \quad J^{\infty}/N \to kX/8k_B T \\ M = 3&: \quad J^{\infty}/N \to kX/12k_B T. \end{aligned} \tag{7.61}$$

These are the linear flux–force relations. It can be proved (1) that the general result is $kX/4Mk_B T$.

The $M = 4$ case is discussed elsewhere (1).

Variation of Ligand Concentration

Suppose, in Fig. 7.27, the concentration of, say, NADH is lowered somewhat (all other concentrations being held constant). This will lower (via μ_{NADH} in Eq. 7.53) the value of X (to X', Fig. 7.27) and also decrease the magnitude of the basic free energy drop for every "α transition" (O → R in enzyme I). The free energy drops for all other transitions remain at $\Delta A'$. The schematic new basic free energy levels are indicated by dotted lines in Fig. 7.27.

To pursue this point still further (in the $M = 3$ case), suppose substrate and/or product concentrations for both end reactions are changed drastically so that for both end reactions $\Delta A'$ in Fig. 7.27 becomes $-2\ \Delta A'$. That is, the end reactions tend now to drive the system in reverse (right to left in Fig. 7.22a). The two "interlocking" reactions, though, retain the same value of $\Delta A'$. Instead of $X = 4\ \Delta A'$, as in Fig. 7.27, we now have $X = -4\ \Delta A' + 2\ \Delta A' = -2\ \Delta A'$. Figure 7.29 shows the new set of basic free energy levels and includes exactly the same transitions as in Fig. 7.27. The (backward) flux in this example would be expected to be relatively

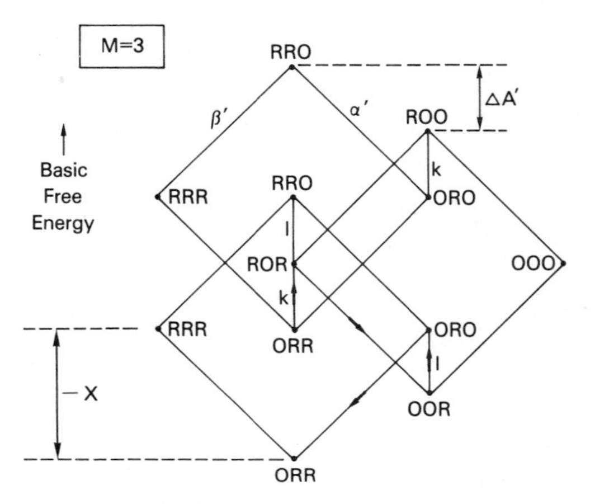

FIG. 7.29 Basic free energy levels for the same system as in Fig. 7.27, but here (because of concentration changes) the α and β free energy differences are altered from $\Delta A'$ to $-2\ \Delta A'$. These reactions now tend to run "backward" (indicated by α' and β' in the figure), driving the whole complex backward, with X now equal to $-2\ \Delta A'$ instead of $+4\ \Delta A'$ (Fig. 7.27).

small because two transitions (OR → RO) in a typical cycle have to go uphill in basic free energy. An example is indicated by the sequence of arrows in Fig. 7.29.

Analogy to Multisite Diffusion Models

It is helpful conceptually and mathematically (see below and reference 7, Section 6) to recognize that there is an exact formal analogy between the above respiratory chain model and the diffusion of a ligand across a membrane, from one bath to another, by means of jumping from site to site along a row of M sites (7,19). A given site may be empty or occupied by one ligand molecule. Ligand adsorption and desorption between each of the baths and sites 1 and M, respectively, are analogous to the two end reactions in Fig. 7.21; the jumping of the ligand from site i to site $i + 1$ is analogous to the interlocking transition RO → OR between enzymes i, $i + 1$; a site occupied by a ligand is equivalent to state R; and an empty site is equivalent to state O. In fact, a ligand molecule is equivalent to a pair of electrons. As is indeed self-evident, the respiratory chain model may thus be regarded as a model for single-file "adsorption, diffusion, and desorption" of electron pairs from one external electron pool (NADH) to another (H_2O; Fig. 7.21).

Recognition of this analogy does not solve the steady-state mathematical problem but it makes equilibrium special cases trivial (even if all rate constants are different). For example, if, say, the $i \rightleftarrows i+1$ transitions mentioned above are blocked by an inhibitor (18), the subsystems 1, 2, ..., i and $i+1$, $i+2$, ..., M will separately come to *equilibrium* at $t = \infty$ (each with its own external pool). All enzymes (sites) are independent at equilibrium, and each will have a "Langmuir adsorption isotherm" of the form (2)

$$P_{\mathrm{R}}^{\infty} = x/(1+x), \qquad P_{\mathrm{O}}^{\infty} = 1/(1+x) \tag{7.62}$$

where it is easy to relate x to rate constant ratios via detailed balance (in general, each site has a different x). It is easy to see (1) that this leads to the well-known "cross-over" phenomenon in the respiratory chain (13, p. 497; 18).

It is important to note that the independence of enzymes at equilibrium, referred to above, does not hold at an arbitrary steady state. Thus, if the rate constant is k for the reaction RO → OR at position r, $r+1$, the rate of this process is in general not equal to $kP_{\mathrm{R}}^{(r)}P_{\mathrm{O}}^{(r+1)}$ (where we omit the superscript ∞). This is the rate at equilibrium and near equilibrium, but not in general. As a simple example, for $M = 2$ and with one-way reactions only, Fig. 7.24 gives $2k/5$ for the correct flux but $kP_{\mathrm{R}}^{(1)}P_{\mathrm{O}}^{(2)} = k(3/5)^2 = 9k/25$.

The use of $kP_{\mathrm{R}}^{(r)}P_{\mathrm{O}}^{(r+1)}$ for the rate of the above process is an approximation (referred to as "mass action" in the footnote on p. 181). The treatment given of single-file diffusion across a membrane in Sections 6C and 6F of Hill and Chen (7) is based on this type of kinetic expression. Hence this treatment is only approximate, though this was not realized at the time reference 7 was written.

I am indebted to Dr. Britton Chance for his stimulating comments on this subject. More realistic special cases are being pursued in collaboration with Dr. Chance and Dr. Don De Vault.

REFERENCES

1. T. L. Hill, *Proc. Nat. Acad. Sci. U.S.* **73**, 4432 (1976); T. L. Hill, *in* "Statistical Mechanics and Statistical Methods in Theory and Application" (U. Landman, ed.). Plenum, New York, 1977.
2. T. L. Hill, "Statistical Thermodynamics." Addison-Wesley, Reading, Massachusetts, 1960.
3. R. Gordon, *J. Chem. Phys.* **49**, 570 (1968).
4. T. L. Hill and Y. Chen, *Proc. Nat. Acad. Sci. U.S.* **70**, 62 (1973).
5. R. Gordon, to be published.
6. T. L. Hill and Y. Chen, *Proc. Nat. Acad. Sci. U.S.* **66**, 607 (1970).
7. T. L. Hill and Y. Chen, *Biophys. J.* **11**, 685 (1971).
8. J.-P. Changeux, J. Thiéry, Y. Tung, and C. Kittel, *Proc. Nat. Acad. Sci. U.S.* **57**, 335 (1967).
9. L. J. Reed and D. J. Cox, *in* "The Enzymes" (P. D. Boyer, ed.), 3rd ed., p. 213. Academic Press, New York, 1970.

10. A. Ginsburg and E. R. Stadtman, *Ann. Rev. Biochem.* **39**, 429 (1970).
11. L. J. Reed, *Accounts Chem. Res.* **7**, 40 (1974).
12. A. J. Stone, *Biochim. Biophys. Acta* **150**, 578 (1968).
13. A. L. Lehninger, "Biochemistry," 2nd ed. Worth, New York, 1975.
14. P. Mitchell, *J. Bioenerg.* **3**, 5 (1973); **4**, 63 (1974).
15. P. D. Boyer, B. Chance, L. Ernster, P. Mitchell, E. Racker, and E. C. Slater, *Ann. Rev. Biochem.* (in press).
16. M. D. Brand, B. Reynafarje, and A. L. Lehninger, *Proc. Nat. Acad. Sci. U.S.* **73**, 437 (1976).
17. D. G. Nicholls, *Eur. J. Biochem.* **50**, 305 (1974).
18. B. Chance and G. R. Williams, *Advan. Enzymol.* **17**, 65 (1956).
19. K. Heckmann, B. Lindemann, and J. Schnakenberg, *Biophys. J.* **12**, 683 (1972).

Appendix 1 | "Reduction" of a Diagram

Suppose we start (1) with, say, a six-state diagram, as in Fig. 1.2c, and with rate constants chosen so that the ensemble of units reaches an equilibrium state at $t = \infty$. Then detailed balance relations will hold between the N_i^e and the rate constants. There will be zero net flux between any two neighboring states in the diagram and zero net flux around any cycle in the diagram (Chapter 2). Now suppose we alter one or more of the above equilibrium set of rate constants so that the ensemble establishes a nonequilibrium steady state at $t = \infty$. In this steady state there will in general be a nonzero net flux between neighboring states and around any cycle (Chapter 2). Thus fluxes can be used to distinguish between equilibrium and a nonequilibrium steady state.

In the six-state diagram it may happen that some of the rate constants are much larger than others so that (a) some states have negligible probabilities (i.e., are transient intermediates) and (b) some neighboring states are in relatively rapid equilibrium with each other. Examples will be considered below. In this case the *effective* number of states in the diagram will be reduced, perhaps to as few as two states. The new reduced effective diagram will also exhibit first-order kinetics, with a new set of effective rate constants (combinations of the original rate constants—see the examples below). If some step in the reduced diagram does *not* follow first-order kinetics, the particular reduction involved (e.g., omission of a presumed transient intermediate) was not justified, and at least one more state must be added to the reduced diagram.

Consider the equilibrium situation again, this time with respect to the reduced diagram with its effective rate constants. Since there are no net fluxes between neighboring pairs of states in the original diagram at equilibrium, there cannot be any net fluxes in inverse processes between neighboring states of the reduced diagram at equilibrium, for the fluxes in the latter case are simply combinations of the (zero) fluxes in the former case. Also, if there were such a net flux between a pair of states in the reduced diagram, there would have to be a cycle with a net flux and this would correspond to a nonequilibrium steady state rather than to equilibrium as specified.

The absence of a net flux for inverse processes between any pair of neighboring states in the reduced diagram requires that a detailed balance relation must hold, for the pair, between the new (combined) equilibrium N_i^e and the effective rate constants. Conclusion: reduction of a diagram does not alter fundamental properties related to detailed balance.

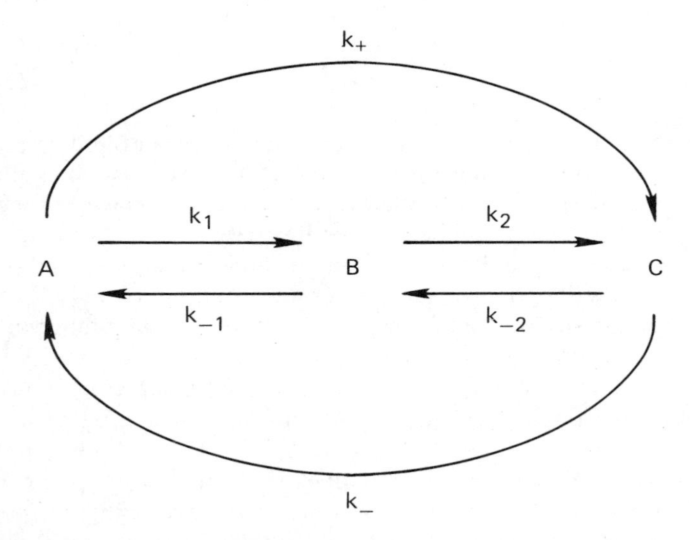

FIG. A1.1 Kinetic scheme for example with a single transient intermediate (B).

Although the above argument is completely general, some examples may be helpful. Also, these examples illustrate relationships between the original and effective rate constants and make explicit the inequality relations needed in order to achieve the assumed reduction in number of states.

Example: Single transient intermediate. Consider three consecutive states in a diagram, A, B, and C, of which B is a transient intermediate (Fig. A1.1). The effective rate constants are k_+ and k_-; the other rate constants shown in the figure are the "original" rate constants. Adding a state before A and another after C does not alter the argument.

We have

$$dp_A/dt = k_{-1}p_B - k_1 p_A, \tag{A1.1}$$

$$dp_B/dt = k_1 p_A + k_{-2}p_C - (k_{-1} + k_2)p_B, \tag{A1.2}$$

$$dp_C/dt = k_2 p_B - k_{-2}p_C, \tag{A1.3}$$

where p_A is the probability of state A, etc., and $p_A + p_B + p_C = 1$. In this example, p_B is assumed to be very small compared to p_A and p_C. Inspection of Fig. A1.1 suggests that this will occur if $k_2 + k_{-1} \gg k_1, k_{-2}$, that is, if either k_2 or k_{-1} is much greater than both k_1 and k_{-2}. In this case, $p_A(t)$ and $p_C(t)$ change on a much slower time scale than $p_B(t)$. In fact, p_B will be able to maintain practically a "steady-state" value appropriate to the relatively slowly changing values of p_A and p_C. In effect, Eq. A1.2 becomes

$$dp_B/dt = \text{const} - (k_{-1} + k_2)p_B.$$

Thus p_B quickly reaches its "steady" value

$$p_B = \frac{\text{const}}{k_{-1} + k_2} = \frac{k_1 p_A(t)}{k_{-1} + k_2} + \frac{k_{-2}p_C(t)}{k_{-1} + k_2} \tag{A1.4}$$

with a time constant $(k_{-1} + k_2)^{-1}$. Note in Eq. A1.4 that $p_B \ll p_A, p_C$, as assumed, if $k_{-1} + k_2 \gg k_1, k_{-2}$. This verifies that this is the appropriate rate constant condition for this special case.

Substitution of Eq. A1.4 for p_B in Eq. A1.1 (or in Eq. A1.3) gives

$$dp_A/dt = k_- p_C - k_+ p_A, \tag{A1.5}$$

where

$$k_+ = \frac{k_1 k_2}{k_{-1} + k_2}, \qquad k_- = \frac{k_{-1}k_{-2}}{k_{-1} + k_2}. \tag{A1.6}$$

As can be seen from Fig. A1.1, Eq. A1.5 is one of the two rate equations for the reduced system A $\rightleftarrows$ C. Equations A1.6 show how the effective constants are related to the original constants.

At equilibrium, we must have (detailed balance)

$$k_1 p_A^e = k_{-1}p_B^e, \qquad k_2 p_B^e = k_{-2}p_C^e$$

or

$$p_A^e/p_C^e = k_{-1}k_{-2}/k_1 k_2. \tag{A1.7}$$

In the reduced system, we must have

$$k_+ p_A^e = k_- p_C^e.$$

This is consistent, as it must be, with Eqs. A1.6 and A1.7.

As a further check, it is easy to show from stochastic theory (2) that the mean first passage time from A to C is

$$\omega_+ = (k_1 + k_2 + k_{-1})/k_1 k_2 \tag{A1.8}$$

If $k_2 + k_{-1} \gg k_1$, this is, in effect, a first-order process with the well-known connection $\omega_+ = 1/k_+$. By symmetry, if $k_2 + k_{-1} \gg k_{-2}$, we have $\omega_- = 1/k_-$ for the process C → A. Thus, we again encounter the requirement, for this special case, that $k_2 + k_{-1} \gg k_1, k_{-2}$.

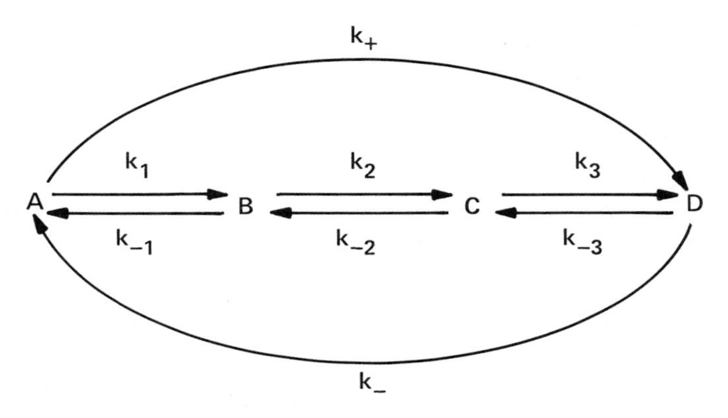

FIG. A1.2 Kinetic scheme for example with two transient intermediates (B, C).

Example: Two transient intermediates. In the system shown in Fig. A1.2, B and C are both transient intermediates, with $p_B \cong 0$ and $p_C \cong 0$. The four rate equations in dp_A/dt, etc., follow from Fig. A1.2. If, as in the above example, and for the same reason, we set the right-hand sides of the equations in dp_B/dt and dp_C/dt equal to zero, and solve for p_B and p_C, we find

$$p_B = \frac{k_1(k_{-2} + k_3)p_A(t) + k_{-2}k_{-3}p_D(t)}{\Phi}, \tag{A1.9}$$

$$p_C = \frac{k_1 k_2 p_A(t) + k_{-3}(k_{-1} + k_2)p_D(t)}{\Phi}, \tag{A1.10}$$

where

$$\Phi = k_{-1}k_{-2} + k_{-1}k_3 + k_2 k_3 .$$

These are the analogues of Eq. A1.4.

Our requirement (in this special case) that $p_B, p_C \ll p_A, p_D$ is met, according to Eqs. A1.9 and A1.10, if

$$\Phi \gg k_1(k_{-2} + k_3),\ k_{-2}k_{-3},\ k_1 k_2,\ k_{-3}(k_{-1} + k_2).$$

These relations are in turn satisfied if

$$k_{-1}, k_{-2} \gg k_1, k_{-3}$$

or if

$$k_{-1}, k_3 \gg k_1, k_{-3} \tag{A1.11}$$

or if

$$k_2, k_3 \gg k_1, k_{-3}.$$

Inspection of Fig. A1.2 confirms that any *one* of these three conditions will guarantee that B and C are present in relatively small amounts.

We now substitute Eq. A1.9 for p_B in the dp_A/dt equation and find

$$dp_A/dt = k_- p_D - k_+ p_A, \tag{A1.12}$$

where

$$k_+ = k_1 k_2 k_3/\Phi, \qquad k_- = k_{-1} k_{-2} k_{-3}/\Phi. \tag{A1.13}$$

Equation A1.12 is one of the two effective (reduced) rate equations.

Equations A1.6 and A1.13 can be extended to any number of transient intermediates.

The detailed balance relation at equilibrium is

$$p_A^e/p_D^e = k_{-1} k_{-2} k_{-3}/k_1 k_2 k_3 = k_-/k_+. \tag{A1.14}$$

The mean first passage time from A to D is found from stochastic theory (2) to be

$$\omega_+ = (k_1 k_2 + k_1 k_3 + k_1 k_{-2} + \Phi)/k_1 k_2 k_3, \tag{A1.15}$$

with a similar expression (by symmetry) for ω_-. Again, any one of the three conditions A1.11 will give

$$\omega_+ = \Phi/k_1 k_2 k_3 = 1/k_+, \qquad \omega_- = \Phi/k_{-1} k_{-2} k_{-3} = 1/k_- \tag{A1.16}$$

for effective first-order processes $A \rightleftarrows D$.

Example: Fast equilibrium. Consider the system shown in Fig. A1.3. The "original" system is the same as in Fig. A1.2 but here there is a "fast equilibrium" between B and C so that in effect there is a single intermediate S = B or C, with $p_S = p_B + p_C$.

Let the fraction of B in the mixture B, C be denoted by θ. Then

$$\theta = p_B/p_S, \qquad 1 - \theta = p_C/p_S.$$

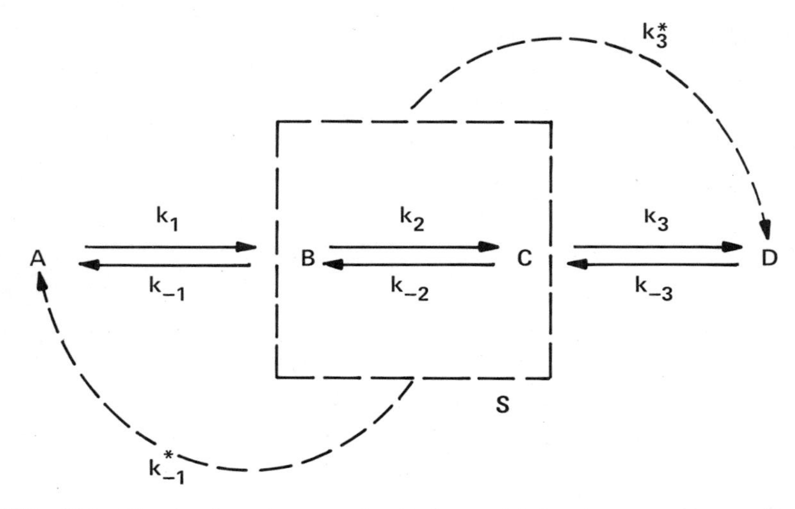

FIG. A1.3 Kinetic scheme for example with fast equilibrium between two states (B, C).

We have four differential equations in dp_A/dt, etc., which follow from Fig. A1.2 or Fig. A1.3. By combining the equations for B and C, we find

$$\begin{aligned} d\theta/dt = {} & k_1(p_A/p_B)\theta(1-\theta) + k_{-2}(1-\theta)^2 \\ & - (k_{-1} + k_2)\theta(1-\theta) \\ & - k_2\theta^2 - k_{-3}(p_D/p_C)\theta(1-\theta) \\ & + (k_{-2} + k_3)\theta(1-\theta). \end{aligned} \tag{A1.17}$$

Since all of the p's are in general of the same magnitude in this example, if we have

$$k_2, k_{-2} \gg k_1, k_{-1}, k_3, k_{-3}, \tag{A1.18}$$

then Eq. A1.17 reduces to

$$\begin{aligned} d\theta/dt &= k_{-2}(1-\theta)^2 - k_2\theta(1-\theta) - k_2\theta^2 + k_{-2}\theta(1-\theta) \\ &= k_{-2}(1-\theta) - k_2\theta. \end{aligned} \tag{A1.19}$$

Here, following a perturbation, θ always quickly restores itself to its essentially constant equilibrium value

$$\theta = k_{-2}/(k_2 + k_{-2}) \tag{A1.20}$$

because the effective time constant involved (Eq. A1.19) is $(k_2 + k_{-2})^{-1}$, which is very small on the time scale determined by k_1^{-1}, k_{-1}^{-1}, k_3^{-1}, and k_{-3}^{-1}. Equation A1.20 can also be written

$$p_B/p_C = \theta/(1-\theta) = k_{-2}/k_2, \tag{A1.21}$$

where p_B and p_C both change slowly with time but maintain this constant equilibrium ratio.

The rate of the transition B → A is

$$k_{-1}p_B = k_{-1}\theta p_S \equiv k^*_{-1}p_S.$$

(The asterisk does not mean "second order" here.) That is, the effective rate constant for S → A in the reduced scheme is

$$k^*_{-1} = k_{-1}\theta = k_{-1}k_{-2}/(k_2 + k_{-2}). \tag{A1.22}$$

Similarly, for S → D,

$$k^*_3 = k_3(1-\theta) = k_2k_3/(k_2 + k_{-2}). \tag{A1.23}$$

At equilibrium, we have the detailed balance relation

$$p^e_A/p^e_B = k_{-1}/k_1 = k^*_{-1}/\theta k_1 \tag{A1.24}$$

or, on multiplying by θ,

$$p^e_A/p^e_S = k^*_{-1}/k_1. \tag{A1.25}$$

There is, of course, an analogous relation involving k^*_3/k_{-3}. Again we see that reduction of the diagram does not affect detailed balance requirements.

It is easy to extend Eqs. A1.22 and A1.23 to equilibria with more than two states. One uses, for example (Fig. A1.4), $k^*_{-1} = k_{-1}(p_B/p_S)$ together with the equilibrium relation between the probability ratio p_B/p_S and the

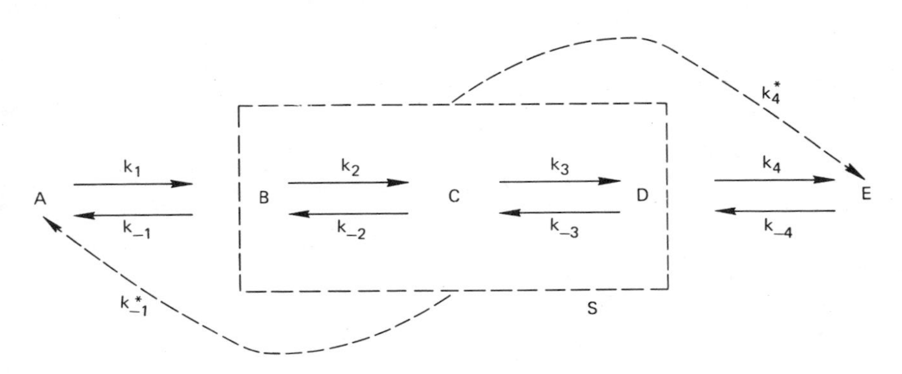

FIG. A1.4 Kinetic scheme for example with fast equilibrium between three states (B, C, D).

rate constants that connect the equilibrated species (B, C, D). Thus, for three states (Fig. A1.4), we find

$$k^*_{-1} = k_{-1}k_{-2}k_{-3}/\Phi', \qquad k^*_4 = k_2 k_3 k_4/\Phi'$$
$$\Phi' = k_2 k_3 + k_{-2}k_{-3} + k_2 k_{-3}\,. \tag{A1.26}$$

REFERENCES

1. T. L. Hill, *Progr. Biophys. Mol. Biol.* **29**, 105 (1975).
2. S. Karlin, "First Course in Stochastic Processes," p. 205. Academic Press, New York, 1966.

Appendix 2 Diagram Solution for the N_i^∞; Flux Diagrams

A prerequisite here is the meaning of *cyclic diagram* and *flux diagram*, as introduced at the beginning of Chapter 2. The problem (1) is posed in Section 1.3.

The solution of the set of linear independent equations in the N_i^∞ is unique, so it suffices for us to *verify* that the diagram solution (Eq. 1.5) does in fact satisfy the linear equations.

The equation $\sum_i N_i^\infty = N$ is obviously satisfied by Eq. 1.5. The remaining task is then to verify that equations of type 1.4 are satisfied. These equations are all homogeneous in the N_i^∞, so for this purpose we can ignore the normalization factor N/Σ in Eq. 1.5.

In an arbitrary diagram, let us examine dN_i/dt at $t = \infty$. Consider a line in the diagram between some other state j, and i. Two terms in $(dN_i/dt)_{t=\infty}$ will be (compare Eq. 1.4).

$$\alpha_{ji} N_j^\infty - \alpha_{ij} N_i^\infty (\equiv J_{ji}^\infty), \tag{A2.1}$$

where α_{ji} and α_{ij} are the rate constants associated with these transitions. This is the net flux into state i along this line. We have to show (using the diagram solution) that the sum of net fluxes into i along all lines emanating from i add up to zero; that is, $(dN_i/dt)_{t=\infty} = 0$.

We begin by examining the diagrammatic interpretation of Eq. A2.1. N_j^∞ is represented by a set of directional diagrams. Multiplication by α_{ji} adds an arrow $(j \rightarrow i)$ to each of these diagrams. N_i^∞ is represented by an equal number of directional diagrams. Both sets of directional diagrams come from the same set of partial diagrams. We consider all pairs of directional diagrams such that one diagram of each pair belongs to N_j^∞ and one belongs to N_i^∞, and both are generated from the same partial diagram. These pairs fall into two classes: (a) there is an arrow between i and j; and (b) there is not an arrow between i and j. These two possibilities are shown schematically in Fig. A2.1. The solid lines and arrows represent the directional diagrams. The short arrows indicate the possibility of other parts of the diagram existing and flowing into i, j or into the path between i and j. The path between i and j is curved to indicate that one or more states may lie on this path. The curved path is unbroken in (b) because one more line (i–j) must make a cycle. The curved path is broken in (a) because a cycle cannot exist in a directional diagram.

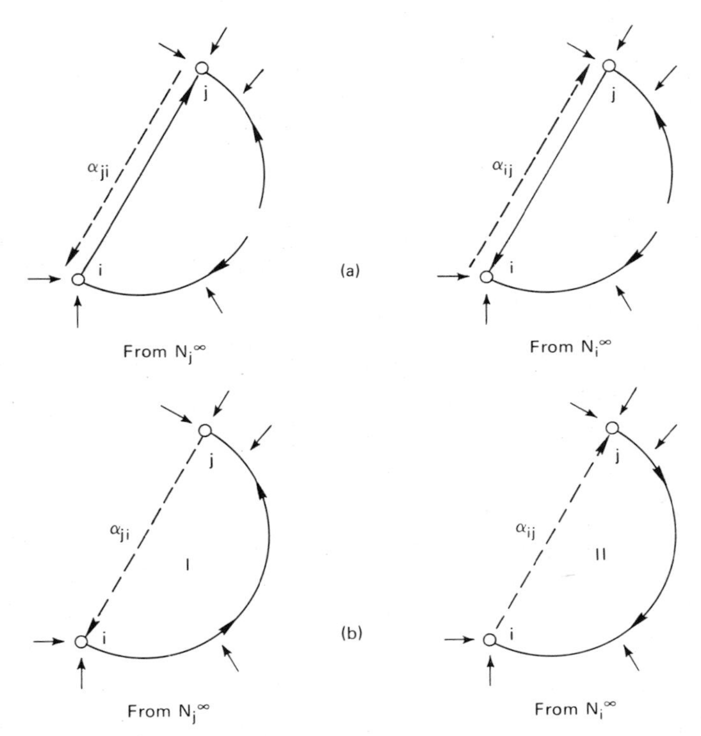

FIG. A2.1 Schematic part of diagram used in proof. See text.

Multiplication by α_{ji} or α_{ij}, as in Eq. A2.1, adds a dashed arrow to the directional diagrams in Fig. A2.1.

After multiplication by α_{ji} and α_{ij}, all pairs of terms of type (a) cancel in forming the difference (Eq. A2.1), because the two directional diagrams in (a), with dotted arrows included, are identical. The dotted arrows produce cycles in (b), but the two *cyclic diagrams* do not cancel in Eq. A2.1 because the cycles have opposite directions (in all other respects, however, the two cyclic diagrams are the same). Thus Eq. A2.1 reduces to a sum of differences of pairs of cyclic diagrams, as in Fig. A2.1b. There will be a pair in this sum for each partial diagram having the i–j line missing.

Now let us turn to the total net flux along all lines into state i. We have to show that this quantity vanishes. There will be a contribution as in Eq. A2.1 from each such line. We have just seen that this contribution, for any one line, is a sum of differences of pairs of cyclic diagrams. Hence the complete expression for $(dN_i/dt)_{t=\infty}$ is also a sum of differences of pairs of cyclic diagrams. We now show that these pairs themselves occur in pairs, and that the pairs of pairs cancel, leading to $(dN_i/dt)_{t=\infty} = 0$.

All the cycles referred to above pass through state i. Furthermore, each cycle makes use of or includes exactly two of the lines emanating from state i. Consider a particular pair of cyclic diagrams, as in Fig. A2.1b, that includes the dotted line i–j and also, say, the line i–l (it may be that $j = l$; that is, there may be two or more lines connecting i and j). This pair is shown in Fig. A2.2a. These cyclic diagrams arise from a certain partial diagram with the line i–j absent but with the line i–l present. Now consider the partial

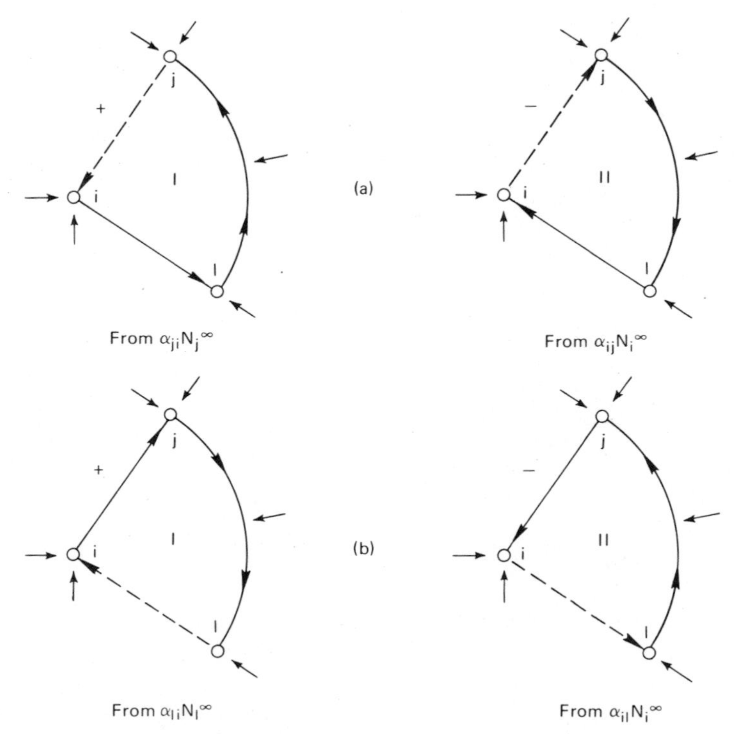

FIG. A2.2 Schematic part of diagram used in proof. See text.

diagram identical with the one just referred to except that the line i–j is present and the line i–l is absent. This partial diagram will contribute a pair of terms to the net flux along the line i–l. This pair is shown in Fig. A2.2b. Each of the four diagrams in Fig. A2.2 contributes to $(dN_i/dt)_{t=\infty}$, and the applicable algebraic signs are included in the figure. It will be seen that I(a) = II(b) and I(b) = II(a). Therefore the four diagrams, including the signs, cancel. This completes the proof of Eq. 1.5.

Transition fluxes from flux diagrams. The above argument leading to Eq. 1.5 (on the calculation of the N_i^∞) contained, in passing, a relation between the *transition flux* J_{ji}^∞ and certain *flux diagrams*. We want to restate this connection more explicitly here.

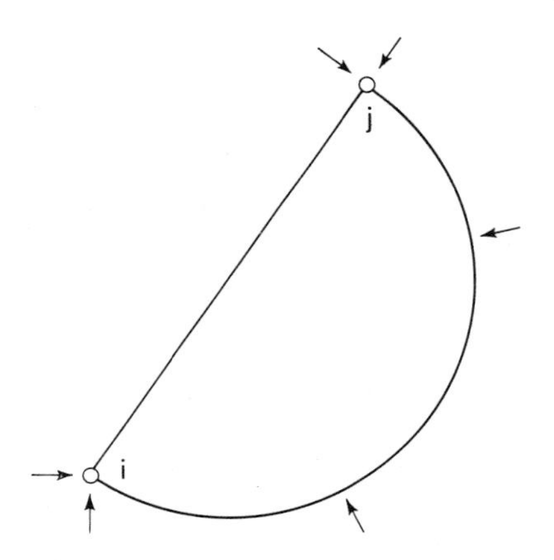

FIG. A2.3 Schematic flux diagram. See text.

The net flux into state i along the line i–j is given by Eq. A2.1, and this in turn is equal to N/Σ in Eq. 1.5 multiplied by a sum of differences of pairs of cyclic diagrams, there being one pair in the sum for each partial diagram having the i–j line missing. We shall write this as

$$J_{ji}^\infty = (N \times \text{sum of flux diagrams for } j \to i)/\Sigma, \qquad \text{(A2.2)}$$

where a *flux diagram* represents a difference between a pair of cyclic diagrams, both derived from the same partial diagram. For example, the flux diagram Fig. A2.3 means the difference I(b) − II(b) in Fig. A2.1. Note that arrows are omitted from the cyclic part of a flux diagram because the one diagram includes both directions around the cycle. It should be recalled that the two cyclic diagrams of a pair, as in Fig. A2.1b, are identical *except* for the direction of the cycle itself. The order of terms in the difference represented by a flux diagram is of course determined by the directionality chosen for the transition flux (i.e., J_{ji}^∞ or J_{ij}^∞).

REFERENCE

1. T. L. Hill, *J. Theoret. Biol.* **10**, 442 (1966).

Appendix 3 Charged Ligand and Membrane Potential

Instead of introducing electrostatic effects into our primary illustrative models of the main text, we treat them here in a relatively simple model. This procedure is possible because electrostatic effects enter into the interpretation or "theory" of the individual rate constants but the formalism developed in this book simply manipulates these rate constants without enquiring into the details of their origin.

Figure A3.1 presents a model (facilitated diffusion) for the transfer of a ligand ion L^z (e.g., K^{+1}) with charge $z\varepsilon$ from bath A at electrostatic potential ψ to bath B at (reference) potential $\psi = 0$. The states 1, 2, and 3 are the same as the first three states in Fig. 1.3a. The local potential at the binding site on E in state 2 is ψ_A; it is ψ_B in state 3. For simplicity, we do not include here any charges on E itself. However, in a more general model, if the local potentials at such charges change values in transitions (by movement of E), these charges and potentials would have to be included in the electrostatic free energy terms introduced below. They would be handled in the same way as $z\varepsilon\psi_A$ and $z\varepsilon\psi_B$, below.

The local potentials $\psi_A(\psi)$ and $\psi_B(\psi)$ are functions of the membrane potential ψ, as are the rate constants $\alpha_{ij}(\psi)$. These functions depend on the potential profile across the membrane and on the location of the binding sites. When $\psi = 0$ (bath A), we assume that $\psi_A = \psi_B = 0$ also.

Let us consider the $\psi = 0$ case first, for reference purposes. The binding rate constants, when $\psi = 0$, are $\alpha_{12}(0) = \alpha^* c_A$ and $\alpha_{13}(0) = \alpha^* c_B$ (i.e., we assume that these are simple diffusion-controlled processes; α^* is the second-order binding rate constant). As usual, we take $c_B = \text{const}$, but c_A is variable. At equilibrium ($\psi = 0$), we have $c_A^e = c_B$ and $\Pi_+ = \Pi_-$:

$$(\alpha^* c_A^e)\,\alpha_{23}(0)\,\alpha_{31}(0) = \alpha_{21}(0)\,\alpha_{32}(0)\,(\alpha^* c_B).$$

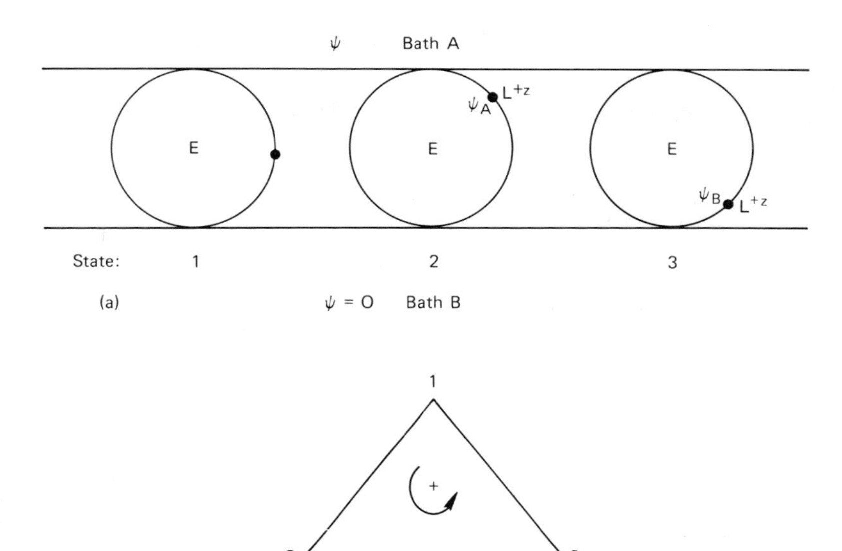

FIG. A3.1 Three-state model for facilitated diffusion of ligand L with charge $+z\varepsilon$. ψ, ψ_A, ψ_B = electrostatic potential values.

Thus there is the required relation between rate constants,

$$\alpha_{23}(0)\,\alpha_{31}(0) = \alpha_{21}(0)\,\alpha_{32}(0). \tag{A3.1}$$

We turn now to the situation with ψ, ψ_A, ψ_B. The binding rate constants are $\alpha_{12}(\psi) = \alpha_{12}^*(\psi)c_A$ and $\alpha_{13}(\psi) = \alpha_{13}^*(\psi)c_B$. The second-order rate constants for binding from the two baths (α_{12}^* and α_{13}^*) may depend on ψ and may now be different because there may be electrostatic barriers, possibly different, to binding at the two locations. Relative to the $\psi = 0$ case, the second-order binding equilibrium constants (and the corresponding standard free energy changes) have to be corrected for the change in electrostatic potential of the ligand ion on binding ($\psi \to \psi_A$ from bath A; $0 \to \psi_B$ from bath B). Thus we have

$$\frac{\alpha_{12}^*(\psi)}{\alpha_{21}(\psi)} = \frac{\alpha^*}{\alpha_{21}(0)}\, e^{(\psi - \psi_A)z\varepsilon/kT} \tag{A3.2}$$

$$\frac{\alpha_{13}^*(\psi)}{\alpha_{31}(\psi)} = \frac{\alpha^*}{\alpha_{31}(0)}\, e^{-\psi_B z\varepsilon/kT}, \tag{A3.3}$$

where the equilibrium constants are expressed as rate constant ratios (Eq. 4.10). Similarly, for the "internal" transitions (where $\psi_A \leftrightarrow \psi_B$),

$$\frac{\alpha_{23}(\psi)}{\alpha_{32}(\psi)} = \frac{\alpha_{23}(0)}{\alpha_{32}(0)} e^{(\psi_A - \psi_B)z\varepsilon/kT}. \tag{A3.4}$$

Equations A3.2 and A3.3 can, of course, also be written (by canceling c_A and c_B)

$$\frac{\alpha_{12}(\psi)}{\alpha_{21}(\psi)} = \frac{\alpha_{12}(0)}{\alpha_{21}(0)} e^{(\psi - \psi_A)z\varepsilon/kT} \tag{A3.5}$$

$$\frac{\alpha_{13}(\psi)}{\alpha_{31}(\psi)} = \frac{\alpha_{13}(0)}{\alpha_{31}(0)} e^{-\psi_B z\varepsilon/kT}. \tag{A3.6}$$

These are the first-order rate constants, related to basic free energy differences (Chapter 4).

Equations A3.2–A3.6 show how *ratios* of inverse rate constants depend on ψ. We need not be completely explicit here but, obviously, if the separate rate constants are to be specified, as required for kinetic calculations, the electrostatic factors must be divided between forward and backward rate constants in each case. This in turn depends on assumptions about the local electrostatic potential in the transition states. Examples are available in the literature (1, 2). Thus, one might use, in Eq. A3.4,

$$\begin{aligned} \alpha_{23}(\psi) &= \alpha_{23}(0)\, e^{(\psi_A - \psi_B)z\varepsilon/2kT} \\ \alpha_{32}(\psi) &= \alpha_{32}(0)\, e^{(\psi_B - \psi_A)z\varepsilon/2kT}. \end{aligned} \tag{A.3.7}$$

This assumes that the potential at the charge $z\varepsilon$, in the transition state for the process $2 \rightleftarrows 3$, is $(\psi_A + \psi_B)/2$.

The thermodynamic force $X(\psi)$ driving the ligand ion in the direction $A \to B$, expressed in terms of the cycle rate constants (Chapter 3), is

$$e^{X(\psi)/kT} = \frac{\Pi_+(\psi)}{\Pi_-(\psi)} = \frac{\alpha_{12}^*(\psi)\, c_A\, \alpha_{23}(\psi)\, \alpha_{31}(\psi)}{\alpha_{21}(\psi)\, \alpha_{32}(\psi)\, \alpha_{13}^*(\psi)\, c_B}. \tag{A3.8}$$

If we substitute Eqs. A3.2–A3.4, and use Eq. A3.1, we find

$$e^{X/kT} = c_A\, e^{\psi z\varepsilon/kT}/c_B. \tag{A3.9}$$

The equilibrium ($X = 0$) value of c_A is

$$c_A^e = c_B\, e^{-\psi z\varepsilon/kT} \tag{A3.10}$$

so that we can also write (compare Section 3.2)

$$e^{X/kT} = c_A/c_A^e. \tag{A3.11}$$

Equation A3.10 is, of course, to be expected since, at equilibrium, the electrochemical potentials of the ligand L in the two baths must be equal:

$$\begin{aligned}\mu_A^e &= \mu_L^0 + kT \ln c_A^e + z\varepsilon\psi \\ &= \mu_B = \mu_L^0 + kT \ln c_B .\end{aligned} \tag{A3.12}$$

At arbitrary c_A,

$$\begin{aligned}\mu_A - \mu_B &= kT \ln c_A + z\varepsilon\psi - kT \ln c_B \\ &= X.\end{aligned} \tag{A3.13}$$

The above discussion confirms that Eq. A3.8 is consistent with ordinary electrochemical thermodynamics.

The steady-state flux is (Eq. 1.20)

$$J^\infty(\psi) = N[e^{X(\psi)/kT} - 1]\Pi_-(\psi)/\Sigma(\psi). \tag{A3.14}$$

The expression [ ] is proportional to $c_A\, e^{\psi z\varepsilon/kT} - c_B$.

It can be seen from the above discussion that electrostatic effects appear not only in the thermodynamics (e.g., in X) but also, in general, in all the rate constants. There is no difficulty in principle in incorporating these effects into a kinetic model, but quite explicit assumptions must be made in each case (1, 2).

Finally, we make explicit the electrostatic contributions to the free energy levels of the states of this system (see Chapter 4). As in Eqs. 4.5 and 4.10, for the $\psi = 0$ case, we have

$$\frac{\alpha^*}{\alpha_{21}(0)} = e^{(A_1 + \mu_L{}^0 - A_2)/kT} \tag{A3.15}$$

$$\frac{\alpha^*}{\alpha_{31}(0)} = e^{(A_1 + \mu_L{}^0 - A_3)/kT} \tag{A3.16}$$

$$\frac{\alpha_{23}(0)}{\alpha_{32}(0)} = e^{(A_2 - A_3)/kT}. \tag{A3.17}$$

For arbitrary ψ, we must make the replacements (in the above equations)

$$\begin{aligned}&A_1 \to A_1 \qquad \text{(no charge on E itself)}\\ &A_2 \to A_2 + z\varepsilon\psi_A\\ &A_3 \to A_3 + z\varepsilon\psi_B\\ &\mu_L^0 \text{ (bath A)} \to \mu_L^0 + z\varepsilon\psi\\ &\mu_L^0 \text{ (bath B)} \to \mu_L^0 \qquad (\psi = 0 \text{ in bath B}).\end{aligned} \tag{A3.18}$$

If E, itself, has charges that undergo changes in their local potential values in transitions, each such charge will make a contribution on the right-hand side of A_1, A_2, A_3 above (like $z\varepsilon\psi_A$ and $z\varepsilon\psi_B$).

If we introduce A3.18 into Eqs. A3.15–A3.17, we recover Eqs. A3.2–A3.4 for arbitrary ψ. That is,

$$\frac{\alpha_{12}^*(\psi)}{\alpha_{21}(\psi)} = e^{(A_1 + \mu_L{}^0 + z\varepsilon\psi - A_2 - z\varepsilon\psi_A)/kT}$$

$$= \frac{\alpha^*}{\alpha_{21}(0)} e^{(\psi - \psi_A)z\varepsilon/kT}, \qquad \text{(A3.19)}$$

etc.

The substitutions A3.18 are also to be used in Table 4.1. For example, for $\Delta A'_{23}$ in the isomeric process $2 \rightarrow 3$,

$$(A_2 - A_3) \rightarrow (A_2 - A_3) + z\varepsilon(\psi_A - \psi_B) \qquad \text{(A3.20)}$$

and, for $\Delta A'_{13}$ in the binding process $1 \rightarrow 3$,

$$(A_1 + \mu_L^0 + kT \ln c_B - A_3) \rightarrow (A_1 + \mu_L^0 + kT \ln c_B - A_3) - z\varepsilon\psi_B\,. \qquad \text{(A3.21)}$$

These changes are consistent with Eqs. A3.4 and A3.6, respectively.

REFERENCES

1. T. L. Hill and Y. Chen, *Proc. Nat. Acad. Sci. U.S.* **66**, 607 (1970).
2. T. L. Hill and Y. Chen, *Biophys. J.* **11**, 685 (1971).

Appendix 4 Some Properties of Single-Cycle Diagrams

Many diagrams contain only a single cycle with n states, where $n = 2, 3, \ldots$. Because of its importance, we give here a summary of selected properties for this special case. Suppose the states occur in the sequence, $1, 2, \ldots, n$, in the plus direction. There are n lines in the cycle. If these are omitted one at a time, we get n partial diagrams. In each one of these, arrows can be introduced leading to any one of the n states, giving n^2 directional diagrams. Examples will be found in Figs. 1.6 ($n = 2$), 1.7 ($n = 3$), and 1.10 ($n = 4$). Then the probability of state i at steady state is (Eq. 1.5)

$$p_i^\infty = \frac{\text{sum of } n \text{ directional diagrams of state } i}{\text{sum of } n^2 \text{ directional diagrams of all states } (\equiv \Sigma)}. \tag{A4.1}$$

Since, at steady state,

$$dN_i^\infty/dt = 0 = J_{i-1,i}^\infty - J_{i,i+1}^\infty, \tag{A4.2}$$

the transition fluxes $J_{i,i+1}^\infty$ ($i = 1, 2, \ldots, n$) are all equal and also equal to the net flux $J^\infty = J_+ - J_-$ around the cycle. In fact (Eq. 2.11)

$$J_+ = N\Pi_+/\Sigma, \qquad J_- = N\Pi_-/\Sigma, \qquad J^\infty = N(\Pi_+ - \Pi_-)/\Sigma, \tag{A4.3}$$

where

$$\Pi_+ = \alpha_{12}\alpha_{23}\cdots\alpha_{n1}, \qquad \Pi_- = \alpha_{21}\alpha_{32}\cdots\alpha_{1n}.$$

The two cycle rate constants are (Eq. 2.14)

$$k_+ = \Pi_+/\Sigma, \qquad k_- = \Pi_-/\Sigma \tag{A4.4}$$

while the cycle probabilities are

$$p_+ = \Pi_+/(\Pi_+ + \Pi_-), \qquad p_- = \Pi_-/(\Pi_+ + \Pi_-). \tag{A4.5}$$

The mean time between completed cycles is (Eq. 2.18)

$$\tau = \Sigma/(\Pi_+ + \Pi_-). \tag{A4.6}$$

The net thermodynamic force X around the cycle in the plus direction (the model need not be specified) is (Chapter 3)

$$e^{X/kT} = \Pi_+/\Pi_- = J_+/J_- . \tag{A4.7}$$

If two or more thermodynamic forces contribute to X, there is complete coupling between the corresponding thermodynamic fluxes. The flux–force relation can be written

$$J^\infty = N(e^{X/kT} - 1)\Pi_-/\Sigma. \tag{A4.8}$$

The rate of entropy production (free energy dissipation), in terms of rate constants, is

$$T\frac{d_iS}{dt} = J^\infty X = NkT\frac{(\Pi_+ - \Pi_-)}{\Sigma}\ln\frac{\Pi_+}{\Pi_-}. \tag{A4.9}$$

The mean time $\tau_{tr}^{(i)}$ in state i, between transitions, is $(\alpha_{i,\,i-1} + \alpha_{i,\,i+1})^{-1}$. Let us denote the numerator in Eq. A4.1 by $\Sigma^{(i)}$, so that $p_i^\infty = \Sigma^{(i)}/\Sigma$ and Σ is the sum of the $\Sigma^{(i)}$. Then the fraction of all transitions that start from state i is

$$f_i \sim p_i^\infty/\tau_{tr}^{(i)} = \Sigma^{(i)}(\alpha_{i,\,i-1} + \alpha_{i,\,i+1})/\Sigma \tag{A4.10}$$

After normalization ($\sum_i f_i = 1$),

$$f_i = \Sigma^{(i)}(\alpha_{i,\,i-1} + \alpha_{i,\,i+1})/S, \tag{A4.11}$$

where S is the sum of the n numerators in Eq. A4.10:

$$S = \sum_i \Sigma^{(i)}(\alpha_{i,\,i-1} + \alpha_{i,\,i+1}). \tag{A4.12}$$

The mean time between transitions is then

$$\tau_{tr} = \sum_i f_i\tau_{tr}^{(i)} = \Sigma/S \tag{A4.13}$$

and the number of transitions per completed cycle is

$$\tau/\tau_{tr} = S/(\Pi_+ + \Pi_-). \tag{A4.14}$$

For example, for Fig. 6.1*c*, $\Pi_+ = 16$, $\Pi_- = 12$, $\Sigma = 127$, $S = 492$, and $\tau/\tau_{tr} = 17.571$ (see p. 136, Section 6.1).

Section 4.2, which is devoted to single-cycle examples and properties, should also be consulted.

Appendix 5 Light Absorbing (and Emitting) Systems

There are well-known and very important energy transducing systems that convert light energy into the free energy of an electrochemical ionic gradient, chemical free energy, etc. Examples are photosynthesis (1) and phototranslocation in the purple membrane of *H. halobium* (2, 3). Such systems fall partly within and partly outside the formalism summarized in this book. The object of this appendix is to use two overly simplified prototypal examples to indicate which parts of the formalism are applicable and which are inapplicable.

We use as our primary example Figs. 4.1 and 4.2a (Section 4.2) for the facilitated diffusion of a ligand L between baths A and B. As usual, for simplicity, we treat the ligand as uncharged. A charged ligand would be handled in essentially the same way (Appendix 3). In Fig. 4.2a, the basic free energy levels correspond to a case in which $c_A > c_B$ and $\mu_A > \mu_B$: there is net steady-state transport of L at the rate $J^\infty > 0$ in the direction A → B as a consequence of the thermodynamic force $X = \mu_A - \mu_B > 0$.

Figure A5.1c shows a modification of this system in which the level A_3 is not above A_4 (as in Fig. 4.2a) but rather is well below it. (Ignore the 3* level to begin with.) That is, 3 → 4 is now a considerable uphill (in basic free energy) transition though 4 → 2, 2 → 1, and 1 → 3 remain downhill as before. In fact, the thermodynamic force in this cycle 34213 (cycle *a* in Fig. A5.1b) is now negative: $c_B > c_A$, $\mu_B > \mu_A$, $X = \mu_A - \mu_B < 0$. Left to itself (i.e., without $h\nu$ and 3*), this system would cycle "backwards" (clockwise in cycle *a* of Fig. A5.1b), on the average, producing a net negative flux J_a—though J_a would be expected to be very small because of the uphill transitions 3 → 1, 1 → 2, and 2 → 4 (in a clockwise cycle) that have to be negotiated before the big free energy drop 4 → 3.

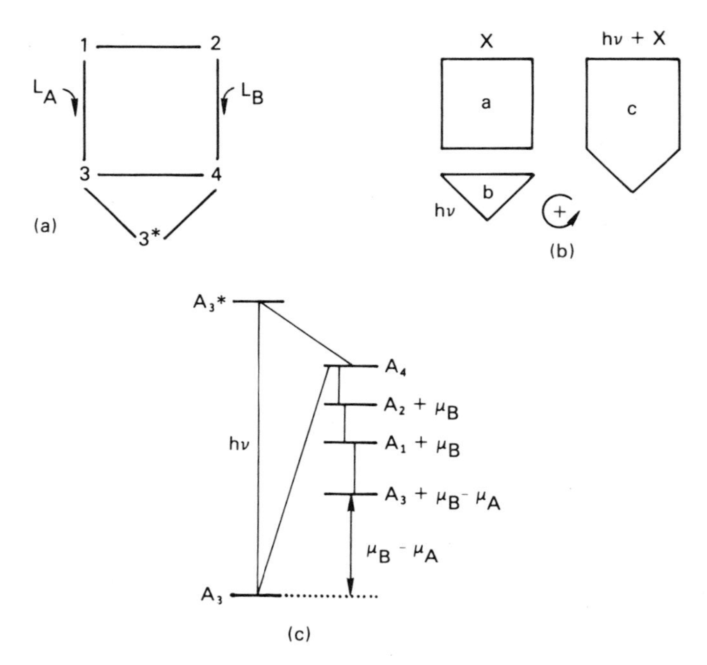

FIG. A5.1 Model that combines facilitated diffusion of ligand L (see Figs. 4.1 and 4.2) with light absorption. State 3* is a new excited electronic state. (a) Diagram. (b) Cycles, with forces indicated (+ direction). (c) Basic free energy levels for macromolecule + ligands (photon excluded).

Now suppose that there exists an excited state, 3*, of state 3, reached from state 3 by absorption of a photon of energy $h\nu$, and suppose further that A_{3*} is above A_4 (as shown in Fig. A5.1c). Transitions (conformational changes—see Fig. 4.1a) between 3* and 4 are assumed possible. State 3* is an excited electronic state, but its lifetime is such that otherwise it has a normal "internal equilibrium" (Section 4.1). Thus, with inclusion of state 3*, the diagram (Fig. A5.1a) expands to include cycles *b* and *c*, as well as cycle *a* (Fig. A5.1b).

It is of course an approximation to show, in Fig. A5.1c, the *free energy* of state 3* above the free energy of state 3 by exactly the photon energy $h\nu$.

If the systems of the ensemble (in the membrane) are exposed to *steady* radiation of frequency ν, of sufficient intensity I, cycle *c* (operating counter-clockwise) will dominate the steady-state proceedings: each traversal of cycle *c* (33*4213) will use part of the photon energy $h\nu$ to transport a molecule of L from bath A to bath B *against* its concentration gradient $c_B > c_A$. Cycle *b* (33*43) is wasteful since a photon is used but no transport is accomplished. The rate constant α_{43} must not be large compared to α_{42} in order to minimize the effect of cycle b and achieve reasonable efficiency. Also, we would expect $\alpha_{3*4} \gg \alpha_{3*3}$ (assuming that fluorescence is weak).

The cycle fluxes J_b and J_c are positive; J_a is negative. J_a is relatively small in magnitude (see above). The rate at which light energy is consumed is $h\nu(J_b + J_c)$; the rate at which transport free energy is produced is $(\mu_B - \mu_A) \times (J_a + J_c)$. The efficiency of the energy → free energy transduction is thus

$$\eta = (-X)(J_a + J_c)/h\nu(J_b + J_c). \tag{A5.1}$$

If J_a and J_b are small compared to J_c, $\eta \cong -X/h\nu$. To the extent that cycles a and b are used, both reduce η (because $J_a < 0$, $J_b > 0$).

The above describes the workings of the model and the transduction. This system simulates phototranslocation coupling (of protons) in the purple membrane of *H. halobium* in a general way.

With the inclusion of state 3*, the diagram in Fig. A5.1a contains twelve first-order rate constants (some of which may be relatively very small or very large). The rate constant (4) $\alpha_{33*} = BI$, where B is a constant and I is the beam intensity, while $\alpha_{3*3} = BI + A$, where A is a constant. B refers to induced absorption or emission, A to spontaneous emission (fluorescence). We assume, in this example, that A is generally large compared to BI. The other ten rate constants are of types already encountered (conformational changes, adsorption, desorption).

As mentioned at the outset, the model just described is oversimplified (for pedagogical purposes). We digress here to discuss modifications that would make it somewhat more realistic, at least for some systems (5). We introduce an additional excited state 3′ (singlet) that has a higher energy than 3* (singlet or triplet). Absorption of a photon of energy $h\nu'$ ($h\nu'$ is greater than $h\nu$ in Fig. A5.1c, and 3′ is above 3*, by, say, 5 kcal mol^{-1}) converts state 3 into state 3′ in an allowed transition with first-order rate constant BI. (The $h\nu$ transition 3 → 3* is forbidden in this modification.) The reverse rate constant (photon emission), for 3′ → 3, is $BI + A$ with $A \gg BI$. But much faster than 3′ → 3 is the radiationless transition (with vibrational relaxation) 3′ → 3*. That is, $\alpha_{3'3*} \gg A$. Thus 3′ is a very short-lived state ($\sim 10^{-12}$ sec) that almost always is thermally deactivated to the lower-lying excited state 3* rather than returning to the ground state by fluorescence (3′ → 3). While the state 3* has a long enough lifetime (say 10^{-7} sec or more) to come to internal equilibrium, at the temperature of the medium (membrane), with respect to vibrational transitions, this is not true of state 3′.

Because 3′ is a transient intermediate with $\alpha_{3'3*} \gg A$, the rate constants for the *effective* transitions 3 ⇄ 3* (via 3′) are (Eqs. A1.6) $\alpha_{33*} = BI$ and $\alpha_{3*3} = A\alpha_{3*3'}/\alpha_{3'3*}$. The latter rate constant (α_{3*3}) is small and of order $A \times e^{-(h\nu' - h\nu)/kT}$ since $\alpha_{3*3'}/\alpha_{3'3*}$ is of order $e^{-(h\nu' - h\nu)/kT}$.

Introduction of the new excited state 3′, as just discussed, does not produce any new cycles in the diagram. However, another realistic possibility, so far not mentioned, would do this. That is, it is quite possible that thermal deactivation of 3* directly to the ground state 3 would compete with $3^* \rightarrow 4$. This would introduce a new path connecting 3* and 3 in the diagram, and three new cycles. Of these, the cycle 33′3*3 would be important and would contribute to inefficiency (along with 33′3*43; i.e., cycle *b*).

Even if we do not include the radiationless transition $3^* \rightarrow 3$ and the new cycle 33′3*3, insertion of the higher excited state 3′ in the model would in any case reduce the efficiency of conversion of radiant energy because the energy $h\nu' - h\nu$ is immediately dissipated as heat in the transition $3' \rightarrow 3^*$. Correspondingly, in Eq. A5.1, $h\nu$ must in this case be replaced by the larger quantity $h\nu'$.

We return now to the original simple model (Fig. A5.1) and ignore the above complications (as well as some others not mentioned).

The twelve first-order rate constants may be used in the equations of Chapters 1, 2, and 6 in the same way as the rate constants belonging to any other diagram. That is, there is nothing unusual or special here about the calculation, from rate constants, of steady-state probabilities of the states of the diagram, cycle fluxes, operational fluxes, fluctuations in these quantities, and the approach to steady state.

The part of the formalism that breaks down (Chapters 3 and 4) is concerned with the photon "thermodynamic force" $h\nu$ and its relation to basic free energy levels, rate constants, entropy production, irreversible thermodynamics, etc.

In order to incorporate photons into the formalism to the maximum extent possible, we now rearrange the basic free energy levels for our simple example as in Fig. A5.2. Here photon absorption is considered *somewhat* analogous to ligand binding (Table 4.1). This is a strictly formal procedure whose only purpose is to show the absorption of a photon in a cycle as a "basic free energy" drop of amount $h\nu$, in analogy with other driving forces we have encountered. However, as will be seen below, in other respects the analogy is not close. The more conventional free energy level system used in Fig. A5.1c can be used if the reader prefers.

Cycle *a* in Fig. A5.2 is conventional. The force $X \equiv \mu_A - \mu_B$ is negative. In cycle *b*, starting at the top left, the first transition (in the counterclockwise direction) is

$$\text{Macromolecule in state 3 } (A_3) + \text{free photon } (h\nu)$$
$$\rightarrow \text{Macromolecule in excited state } 3^* \ (A_{3*}). \qquad \text{(A5.2)}$$

This transition occurs with no change in basic free energy (in the approximation we are using, mentioned above). This is because, although the free energy levels in Fig. A5.1c refer (conventionally) to macromolecule + ligands, in Fig. A5.2 the levels refer to macromolecule + ligands + *photon*. In one circuit of cycle *b* (33*43), in Fig. A5.2, the basic free energy drop is $h\nu$, the photon "force." All of this energy is dissipated as heat.

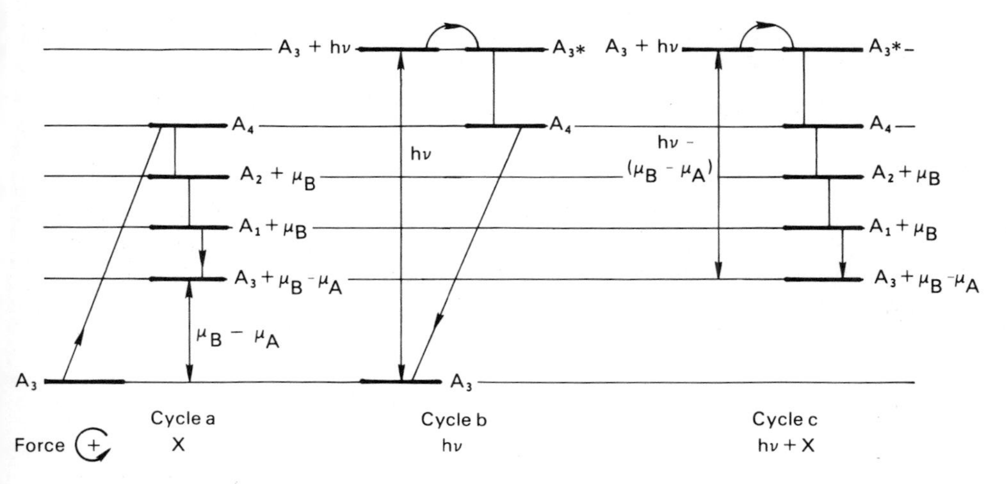

FIG. A5.2 Basic free energy levels and forces (+ direction), for cycles of Fig. A5.1b, on the basis of macromolecule + ligands + photon. Unidirectional arrows in each cycle show + direction of transitions. For each cycle, the sets of free energy levels are repeated above and below the one set shown.

Similarly, in one counterclockwise circuit of cycle *c*, 33*4213, the basic free energy drop (Fig. A5.2) is $h\nu + X$, the net force for this cycle. Transduction occurs here. Part of $h\nu$ appears as transport free energy $-X$; the rest, $h\nu + X$, is dissipated as heat.

Up to this point, photons seem to qualify as a "ligand," but the analogy cannot be pushed any further. Photons are not material particles that are conserved. Also, a ligand has well-defined thermodynamic properties (e.g., a chemical potential) whereas a beam of photons of frequency ν and arbitrary intensity I is decidedly nonthermodynamic. Only photons in blackbody radiation have well-defined thermodynamic functions.

The fundamental relation (Eq. 4.14)

$$\alpha_{ij}/\alpha_{ji} = \exp(\Delta A'_{ij}/kT) \tag{A5.3}$$

applies to all transitions in Fig. A5.2 *except* $3 \rightleftarrows 3^*$ (in cycles *b* and *c*). This follows because the ratio $\alpha_{33*}/\alpha_{3*3}$ is equal to $BI/(BI + A) \cong BI/A$, but $\Delta A'_{33*} \equiv 0$. That is, $\Delta A'_{33*}$ is a constant but $\alpha_{33*}/\alpha_{3*3}$ depends on I. Also, the ratio $\alpha_{33*}/\alpha_{3*3}$, for the photon beam at arbitrary I, has nothing to do with detailed balance at equilibrium (because this is not blackbody radiation*), whereas Eq. A5.3 is based on detailed balance (see Eqs. 4.3–4.8).

* In a blackbody cavity at equilibrium, with no external beam, I is not a variable but has a value *prescribed* by ν and T (4) such that $\alpha_{33*}/\alpha_{3*3} = e^{-h\nu/kT}$.

The failure of $\alpha_{33*}/\alpha_{3*3}$ (photon absorption and emission) to obey Eq. A5.3 has several ramifications:

(a) For cycles b and c we have (compare Eqs. 3.6, 3.8, and 3.10)

$$\Pi_{\kappa+}/\Pi_{\kappa-} = J_{\kappa+}/J_{\kappa-} \qquad (\kappa = b, c) \tag{A5.4}$$

as usual, since this is a purely kinetic result. But Π_{b+}/Π_{b-} is *not* equal to $e^{h\nu/kT}$ and Π_{c+}/Π_{c-} is *not* equal to $e^{(h\nu+X)/kT}$. This is obvious since both Π quotients depend on intensity I whereas $h\nu$ and X do not. Even if I becomes very small, the "force" $h\nu$ remains constant.

In fact, as is obvious from Fig. A5.2,

$$\begin{aligned} \alpha_{3*4}\alpha_{43}/\alpha_{43*}\alpha_{34} &= e^{h\nu/kT} \\ \alpha_{3*4}\alpha_{42}\alpha_{21}\alpha_{13}/\alpha_{43*}\alpha_{24}\alpha_{12}\alpha_{31} &= e^{(h\nu+X)/kT}. \end{aligned} \tag{A5.5}$$

These quotients do not include α_{33*} and α_{3*3} at all.

(b) Clearly, the gradual approach to equilibrium as a limiting steady state (achieved by varying rate constants) and the application of near-equilibrium irreversible thermodynamics (reciprocal relations) have to be abandoned if one of the "thermodynamic forces" ($h\nu$ = const) is provided by a monochromatic beam of photons. Of course if the beam $I \equiv 0$, the rest of the system (with conventional forces) could approach equilibrium gradually in the usual way, i.e., by varying ligand, substrate, etc., concentrations.

(c) The rate of entropy production or free energy dissipation in ensemble + baths is $(J_a + J_c)X$, a *negative* quantity. This refers to ligand transport only. If we include the photons, the net rate of loss of energy + free energy is

$$h\nu(J_b + J_c) + X(J_a + J_c) > 0,$$

but this is not a "thermodynamic" quantity—since the photon beam is not a thermodynamic system.

(d) Gross free energy levels can be defined as before (Chapter 4), as this involves only addition of $kT \ln p_i$ to A_i' for state i. But Eq. 4.27 is inapplicable to the photon transitions because Eq. A5.3 does not hold. Thus J_{ij} (Eq. 4.26) need not have the same sign as $\Delta\mu_{ij}' = kT \ln(p_i/p_j)$ for these transitions. Also, Eq. 4.28 is no longer applicable when photon transitions are included. Hence the gross levels, at least as a complete "package" for a diagram, lose their interest to a considerable extent.

In summary, the diagram method for steady-state probabilities, cycle fluxes, and fluctuations (Chapters 1, 2, and 6) can be applied without change to photon-absorbing (and photon-emitting) systems. But some important relations between diagram rate constants, on the one hand, and thermodynamics, on the other (Chapters 3 and 4), are no longer valid. This is due, in essence, to the facts that photons are not material particles and that the photon beam is nonthermodynamic.

Example with chemical reaction. Consider an enzyme E that catalyzes S ⇄ P. We define the chemical force as $X = \mu_S - \mu_P$. We assume, in this example (Fig. A5.3), that X is negative (the thermodynamic tendency is for P → S) and that the basic free energy levels (without E* and $h\nu$) are arranged as in cycle a of Fig. A5.4. The flux J_a is negative and would be expected to be small in magnitude because of the uphill transitions E → EP and EP → ES (in the clockwise direction of cycle a).

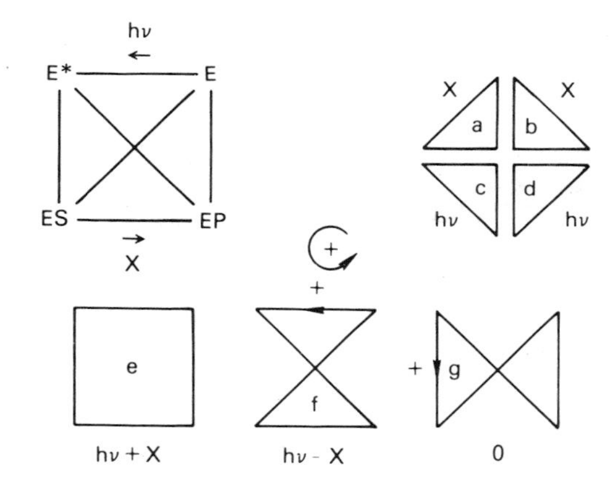

FIG. A5.3 Diagram, cycles, and forces (+ direction) for enzyme E with excited electronic state E*, attainable from E by light absorption. The positive direction of $X \equiv \mu_S - \mu_P$ (by definition) is indicated on the diagram. In the case in Fig. A5.4 and the text, X is negative. The + direction for cycles f and g is indicated by an arrow on the cycles; for the other cycles, + = counterclockwise.

We now assume that E has an excited electronic state E*, of lifetime, say 10^{-6} sec, accessible from E by absorption of a photon of energy $h\nu$, and further that E* binds S strongly (and sufficiently rapidly): the transition E* → ES is downhill in basic free energy whereas E → ES is uphill (Fig. A5.4). (Perhaps a more realistic model would use photon absorption in the step ES → E*S, i.e., *after* binding of S.)

The diagram with E* included is shown in Fig. A5.3, together with the cycles and the force associated with each cycle (plus direction). The hypothetical basic free energy levels for each cycle are given in Fig. A5.4. The most important cycle is presumed to be cycle e. One counterclockwise circuit around this cycle uses part of the photon energy $h\nu$ to transform one

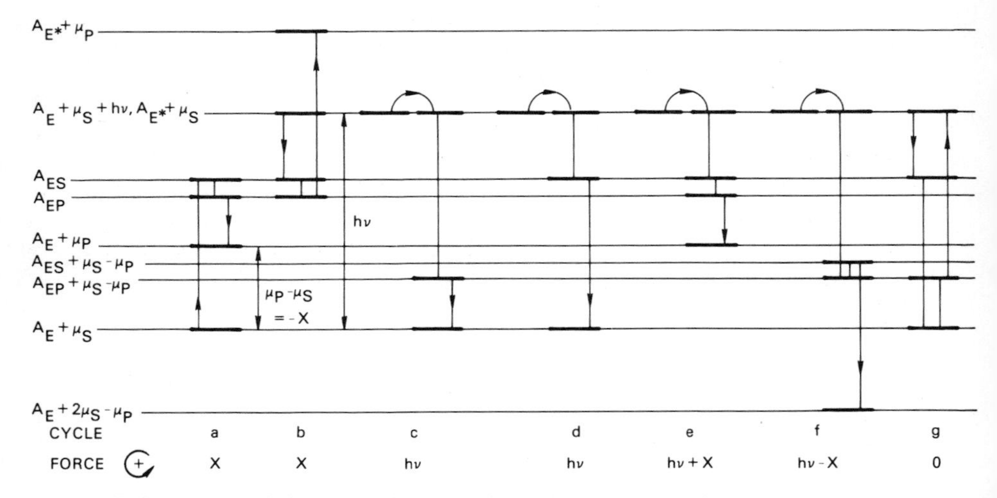

FIG. A5.4 Basic free energy levels and forces, in text example, for the cycles of Fig. A5.3, on the basis of macromolecule + ligands + photon. Unidirectional arrows in each cycle show + direction of transitions. For each cycle, the sets of free energy levels are repeated above and below the one set shown.

molecule of S into P despite the fact that $\mu_P > \mu_S$. Thus we have *energy → free energy* transduction, in cycle *e*, in which the photon energy $h\nu$ is partially converted into chemical free energy $\mu_P - \mu_S$.

Cycle *b*, like cycle *a*, would also have a negative but small flux (because of uphill transitions EP → ES and ES → E*). Cycles *c* and *d* are wasteful because a photon is used but no chemical reaction occurs. Cycle *f* runs the chemical reaction in the downhill direction P → S and hence is counterproductive (if S → P is the objective). Cycle *g* has zero flux and force.

The efficiency of the conversion of photon energy $h\nu$ into chemical free energy $\mu_P - \mu_S$ is

$$\eta = \frac{(-X)(J_a + J_b + J_e - J_f)}{h\nu(J_c + J_d + J_e + J_f)}. \tag{A5.6}$$

In this equation, J_a and J_b are negative and the other fluxes positive. Thus all cycle fluxes other than J_e tend to diminish the efficiency. If only J_e is important, $\eta \cong -X/h\nu$.

If the basic free energy levels are altered so that $\mu_P - \mu_S$ is larger than $h\nu$, rather than smaller, it is possible for the system to run "backwards" (with $I = 0$ and $\alpha_{EE*} = 0$): part of the chemical free energy $\mu_P - \mu_S$ is converted into spontaneously emitted radiant energy $h\nu$ by the transition E* → E, as in

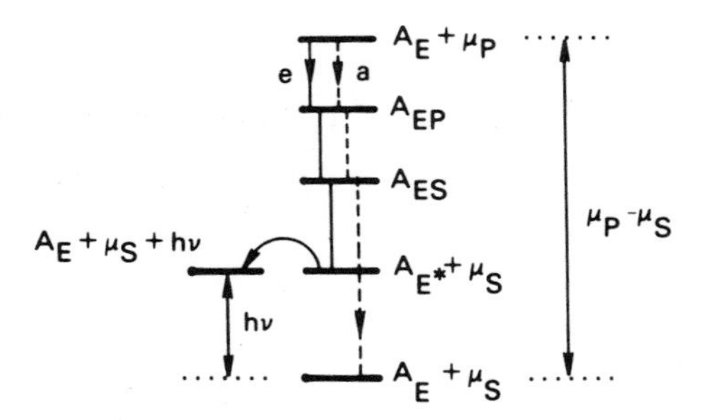

FIG. A5.5 The model of Fig. A5.3 with modified basic free energy levels (partial set) for a case in which chemical free energy ($\mu_P - \mu_S$) is partially transformed into emitted light ($h\nu$). Transitions in only two cycles (a and e) are shown. Unidirectional arrows are in clockwise (−) direction in this case.

bioluminescence. This is illustrated in Fig. A5.5, where only cycles a and e are included (for simplicity). Cycle a (clockwise) completely dissipates the free energy $\mu_P - \mu_S$ but cycle e (clockwise) converts part of $\mu_P - \mu_S$ into $h\nu$. The remainder, $(\mu_P - \mu_S) - h\nu$, is dissipated.

REFERENCES

1. H. T. Witt, *Q. Rev. Biophys.* **4**, 365 (1971).
2. D. Oesterhelt and W. Stoeckenius, *Proc. Nat. Acad. Sci. U.S.* **70**, 2853 (1973).
3. A. Danon and W. Stoeckenius, *Proc. Nat. Acad. Sci. U.S.* **71**, 1234 (1974).
4. G. M. Barrow, "Molecular Spectroscopy," pp. 70–72, 307–311. McGraw-Hill, New York, 1962.
5. R. R. Birge, K. Schulten, and M. Karplus, *Chem. Phys. Lett.* **31**, 451 (1975).

Appendix 6 Basic and Gross Free Energy Levels in a Simple Special Case

Some readers may find it helpful to have a more thorough discussion of free energy values and related topics, as introduced in Chapter 4, for a simple special case. The model used is the same as in Section 6.3 and Fig. 6.6.

We treat an ensemble of N equivalent, independent, and distinguishable macromolecular systems or units, each of which can be in one or the other of two isomeric states, or conformations, labeled 1 and 2. Thus, $N = N_1 + N_2$. For example, the macromolecules might be immobilized in a membrane and are sufficiently separated so that they do not interact with each other. The first-order rate constants for the conformational change are α_{12} and α_{21}. When the ensemble is at equilibrium, detailed balance gives us

$$\alpha_{12} p_1^e = \alpha_{21} p_2^e, \qquad p_1^e / p_2^e = \alpha_{21} / \alpha_{12} \equiv K, \tag{A6.1}$$

where $p_1 = N_1/N$ is the fraction of systems in state 1 and $p_2 = 1 - p_1 = N_2/N$.

The partition functions and free energies of *individual* units are

$$A_1 = -kT \ln Q_1 \qquad \text{and} \qquad A_2 = -kT \ln Q_2 . \tag{A6.2}$$

We assume here (with very small error) that the macromolecules have negligible pV_i terms so that A_i and Q_i ($i = 1, 2$) are functions of T only. Otherwise we would use $G_i(p, T)$ and the corresponding partition function $\Delta_i(p, T)$ (1, 2). Other thermodynamic relations for individual units are

$$\begin{aligned} A_i &= E_i - TS_i, & dA_i &= -S_i\, dT \\ S_i &= k \ln Q_i + kT \frac{d \ln Q_i}{dT}, & E_i &= kT^2 \frac{d \ln Q_i}{dT}. \end{aligned} \tag{A6.3}$$

To begin with, consider the ensemble at some particular instantaneous composition N_1, N_2. Then the partition function and free energy of the entire ensemble are

$$Q = \frac{(N_1 + N_2)!\,Q_1^{N_1} Q_2^{N_2}}{N_1!\,N_2!} \qquad \text{and} \qquad A = -kT \ln Q. \tag{A6.4}$$

Formally, this is the partition function of an ideal binary solid solution (1). The factorial expression takes care of the configurational degeneracy. Other thermodynamic properties of the ensemble are

$$\begin{aligned} dG = dA &= -S\,dT + \mu_1\,dN_1 + \mu_2\,dN_2 \\ G = A &= E - TS = \mu_1 N_1 + \mu_2 N_2 \end{aligned} \tag{A6.5}$$

$$\begin{aligned} \mu_i = (\partial A/\partial N_i)_{N_j,\,T} &= -kT(\partial \ln Q/\partial N_i)_{N_j,\,T} \\ &= A_i + kT \ln p_i \qquad (i = 1, 2). \end{aligned} \tag{A6.6}$$

Also, using Eqs. A6.3,

$$\begin{aligned} E &= N_1 E_1 + N_2 E_2 \\ S &= N_1 S_1 + N_2 S_2 + S_c \\ A &= N_1 A_1 + N_2 A_2 - TS_c, \end{aligned} \tag{A6.7}$$

where S_c is the configurational entropy, or "entropy of mixing,"

$$\begin{aligned} S_c = k \ln \frac{(N_1 + N_2)!}{N_1!\,N_2!} &= -N_1 k \ln p_1 - N_2 k \ln p_1 \\ &= -Nk(p_1 \ln p_1 + p_2 \ln p_2). \end{aligned} \tag{A6.8}$$

Note from Eq. A6.7c that μ_i can also be written as

$$\mu_i = \partial A/\partial N_i = A_i - T\,\partial S_c/\partial N_i. \tag{A6.9}$$

That is, in Eq. A6.6,

$$kT \ln p_i = -T\,\partial S_c/\partial N_i. \tag{A6.10}$$

This makes clear the configurational entropy origin of the ln p_i term in μ_i. This term is a property of the ensemble of systems and in general is not defined for any single system (macromolecule) of the ensemble. See, however, the discussion at the end of this appendix (pp. 225–226) on the substitution of stochastic and time averages for ensemble averages.

Since only isomeric states are involved in this example, the "basic free energy" of a macromolecule in state i (Chapter 4) is A_i while the "gross free energy" is $\mu_i = A_i + kT \ln p_i$. The basic free energy is a property of an *individual unit* in state i (Eqs. A6.2 and A6.3) while μ_i is an *ensemble* property of units in state i.

If all units in the ensemble are in state i ($N_i = N$, $p_i = 1$), then $A = N_i A_i$ (Eq. A6.7c) and the free energy per molecule is $A_i = A/N_i = \mu_i$ (Eq. A6.5b). Thus A_i in Eq. A6.6 may also be considered to be the standard chemical potential of state i with standard state chosen as $p_i = 1$ (all units in state i).

Kinetic behavior of the ensemble. Consider, now, the instantaneous *kinetic* behavior of an ensemble at an arbitrary composition N_1, N_2. Let us suppose, for concreteness, that at this composition we have $\mu_2 > \mu_1$. Now let there be a small amount dN_2 of spontaneous reaction at constant T (with $dN_1 = -dN_2$, since $N = \text{const}$). Equation A6.5a then becomes

$$dA = \mu_1 \, dN_1 + \mu_2 \, dN_2 = (\mu_2 - \mu_1) \, dN_2 . \tag{A6.11}$$

According to the second law of thermodynamics, dA must be negative here. Hence dN_2 must also be negative. That is, there is net reaction in the direction $2 \to 1$.

We digress to note that equilibrium is the special case $dA = 0$. Thus $\mu_2 = \mu_1$ at equilibrium. From Eqs. A6.1 and A6.6,

$$p_1^e / p_2^e = \alpha_{21}/\alpha_{12} = K = e^{(A_2 - A_1)/kT}. \tag{A6.12}$$

This establishes a connection between α_{21}/α_{12} and $A_2 - A_1$. Since both of these quantities are intrinsic properties of an *individual* unit, the relation between them must hold whatever the state of the ensemble (i.e., whether the ensemble is at equilibrium or not). In fact, this relation must hold for an individual unit even if there is no ensemble ($N = 1$).

We return now to our kinetic discussion. The quantity $\mu_2 - \mu_1$ in Eq. A6.11 is the thermodynamic force that drives the ensemble toward equilibrium. Alternate expressions for $\mu_2 - \mu_1$ are

$$\mu_2 - \mu_1 = (A_2 - A_1) + kT \ln(p_2/p_1) \tag{A6.13}$$

$$= kT \ln(\alpha_{21} p_2/\alpha_{12} p_1) \tag{A6.14}$$

$$= kT \ln[(p_2/p_1)/(p_2^e/p_1^e)]. \tag{A6.15}$$

The flux, at the instant being considered, is defined as

$$dN_1/dt = N(\alpha_{21} p_2 - \alpha_{12} p_1). \tag{A6.16}$$

On comparing Eqs. A6.14 and A6.16, we see that $\mu_2 - \mu_1$ and dN_1/dt always have the same sign: e.g., when $\mu_2 > \mu_1$, as above, there is net reaction in the direction $2 \to 1$. Note that, from Eq. A6.11, $-dA/dt = \text{force} \times \text{flux}$ (Eq. 4.28). This is the rate of free energy dissipation or entropy production.

We have already seen in Eq. A6.11 that, if a small amount of reaction takes place in the ensemble, the quantity $\mu_2 - \mu_1$ (the gross free energy difference for the process $2 \rightarrow 1$) has the physical significance of the change in the ensemble free energy A ($= G$) per molecule reacting, i.e., $(\partial A/\partial N_2)_{T,N}$. Correspondingly, it is easy to see that, in the same process, $A_2 - A_1$ (the basic free energy difference) has the physical significance—at the ensemble level—of the change in $A + TS_c$, or in $N_1 A_1 + N_2 A_2$, per molecule reacting. $N_1 A_1 + N_2 A_2$ is the "intrinsic" free energy of the ensemble—i.e., the free energy with the configurational entropy excluded (Eq. A6.7c).

The force and flux, at an arbitrary time, are ensemble properties that depend on the statistics of many (N) units. Each individual unit in the ensemble makes independent and random transitions, completely oblivious of the rest of the ensemble (or its state), with transition probabilities α_{12} and α_{21}. But a large collection of N units exhibits a definite *statistical* trend toward equilibrium at a definite statistical rate—if the instantaneous values of p_1 and p_2 in the ensemble do not correspond to the equilibrium values p_1^e and p_2^e. The quantity $A_2 - A_1$ (the basic free energy difference) is related via α_{21}/α_{12} to the stochastic behavior of each *individual* unit while the quantity $\mu_2 - \mu_1$ (the gross free energy difference) relates to the statistical behavior of the *ensemble* taken as a whole.

In Eq. A6.13, the term $A_2 - A_1$ makes an *invariant* contribution to $\mu_2 - \mu_1$, but the *time-dependent* term $kT \ln(p_2/p_1)$ depends on the state of the whole ensemble. For individual units, there is an intrinsic or "basic" drive $A_2 - A_1$ in the direction $2 \rightarrow 1$ if $A_2 > A_1$ or in the direction $1 \rightarrow 2$ if $A_1 > A_2$. But the statistical or "gross" drive $\mu_2 - \mu_1$ in the whole ensemble depends not only on $A_2 - A_1$ but also on $kT \ln(p_2/p_1)$; as a result, $\mu_2 - \mu_1$ and $A_2 - A_1$ could even have opposite signs. At equilibrium, $kT \ln(p_2^e/p_1^e)$ exactly neutralizes $A_2 - A_1$ so that $\mu_2 - \mu_1 = 0$.

Kinetics of a single unit. Suppose we consider only a single unit but follow its stochastic history repeatedly (N times), starting each "run" at $t = 0$. If we then average over the N runs at each t, the equations of the preceding subsection would still apply at any t provided that the state (1 or 2) of the subunit at $t = 0$ is assigned, in the N runs, in the same ratio as the initial ($t = 0$) value of N_1/N_2 chosen for a single run of the whole ensemble (N units). Ensemble averages are then replaced by stochastic averages.

But if we follow a single unit over an essentially infinite length of time (without repetitions), time averaging will produce only equilibrium properties. There is no "trend toward" or approach to equilibrium in this case. An example: whenever the unit reaches state 1 (following a transition), the mean time spent in state 1 before the next transition is α_{12}^{-1}. Similarly, for state 2, this mean time is α_{21}^{-1}. Thus the ratio of time spent in the two states (at equilibrium) is

$$p_1^e/p_2^e = \alpha_{12}^{-1}/\alpha_{21}^{-1} = \alpha_{21}/\alpha_{12}, \tag{A6.17}$$

which agrees with Eq. A6.1. That is, the long-time average here, for one unit, is the same as the (equilibrium) ensemble average.

Single unit at steady state. A generalization of the preceding paragraph is to examine the stochastic behavior of a single unit, over a very long time, when the kinetic diagram (Chapter 1) is arbitrarily complicated and when the rate constants of the diagram have values such that a nonequilibrium steady state replaces the equilibrium state referred to above. We have, in fact, already made such an examination, for several examples, in Section 6.1.

The properties found from the behavior of a single unit over a long period of time all relate only to the steady state of an ensemble of such units. There is no way to deduce anything, from this single long-time record, about the *approach* of an ensemble to steady state. But it *is* possible (Section 6.1) to deduce from this one record the state probabilities, cycle fluxes, thermodynamic fluxes, gross free energy levels, thermodynamic forces, etc., of an *ensemble* at *steady state.* That is, long-time averages for one unit are equal to steady-state ensemble averages.

Single cross-bridge in a steady isotonic contraction. For completeness, we should also discuss this more complicated system from the above point of view. See Chapter 5 and Section 6.2 for the necessary background.

We can use a very large number N of passes of an actin site (at velocity v) past one cross-bridge in lieu of an ensemble of N cross-bridges (with a subensemble at each x value). From the stochastic record of each of the N passes, averages can be taken at each x value (rather than at each time t, as above). Thus, at each x, one can calculate (for the velocity v) state probabilities, gross free energy levels, fluxes, etc. Thus, for steady isotonic contractions, ensemble averages can be replaced by stochastic averages.

But long-time averaging of a single cross-bridge is not generally relevant for this problem because t must be replaced by x as an independent variable and x-averaging is carried out over a finite interval only (because of periodicity in x). The exception is steady isometric contraction ($v = 0$) where we can use long-time averaging (as above) of the stochastic behavior of a cross-bridge at a fixed x. But, to obtain operational quantities, this must be supplemented by x-averaging as a final step.

REFERENCES

1. T. L. Hill, "Statistical Thermodynamics," pp. 30, 371. Addison-Wesley, Reading, Massachusetts, 1960.
2. T. L. Hill, *Progr. Biophys. Mol. Biol.* **28**, 267 (1974). See p. 272.

Index

N

O

P

R

S

T

U

V

W

A
B 7
C 8
D 9
E 0
F 1
G 2
H 3
I 4
J 5